KU-492-888

Darken Ship
The Master Mariner book two

The son of a distinguished surgeon, Nicholas Monsarrat was born in Liverpool in 1910 and was educated at Winchester and Trinity College, Cambridge. His first book to attract attention was the largely autobiographical *This Is the Schoolroom*, published in 1939. On the outbreak of war he joined the RNVR, serving mainly with corvettes: his war-time experiences are vividly described in *The Three Corvettes* and *Depends What You Mean by Love*. In 1946 he became director of the UK Information Service in Johannesburg and subsequently in Ottawa. His most famous book, *The Cruel Sea*, published in 1951, is one of the most successful stories of all time and was made into a film starring Jack Hawkins. Other famous novels include: *The Tribe that Lost its Head*, *Richer than All His Tribe*, *The Kapillan of Malta*, *The Story of Esther Costello*, *The White Rajah* and *The Pillow Fight*.

His last novel, *The Master Mariner*, was the result of many years of research and the culmination of a long-held ambition, and consists of Book 1, *Running Proud*, and Book 2, *Darken Ship*. Nicholas Monsarrat died in 1979.

Also by Nicholas Monsarrat
in Pan Books

The Tribe That Lost Its Head
The Ship That Died of Shame
The Nylon Pirates
The White Rajah
Richer Than All His Tribe
The Pillow Fight
The Kapillan of Malta

The Master Mariner
book one Running Proud

Nicholas Monsarrat

Darken Ship
**The Master Mariner
book two**

with a biographical introduction by
Ann Monsarrat

Pan Books London and Sydney

First published 1980 by Cassell Ltd
This edition published 1981 by Pan Books Ltd,
Cavaye Place, London SW10 9PG
© The Estate of the late Nicholas Monsarrat 1980
ISBN 0 330 26553 9
Printed and bound in Great Britain by Collins, Glasgow

This book is sold subject to the condition that it
shall not, by way of trade or otherwise, be lent, re-sold,
hired out or otherwise circulated without the publisher's prior
consent in any form of binding or cover other than that in which
it is published and without a similar condition including this
condition being imposed on the subsequent purchaser

'Let me end my days somewhere where the tide comes in and out: leaving its tribute, its riches, taking nothing. Giving all the time: pieces of wood, pieces of eight: seaweed for the land, logs for the fire, sea-shells for pleasure, skeletons for sadness.'

Darken Ship
The Master Mariner book two

Introduction

Ann Monsarrat

The Master Mariner was, on my husband's own admission, a grand design with an unworthy conception. It began in 1956, five years after the publication of *The Cruel Sea*. He was living then in Canada and had just become, for the first time since before the war, a full-time writer, burning the boats of security behind him by giving up his diplomatic job of Information Officer.

At forty-six, with his twelfth book, *The Cruel Sea*, still notching up record sales, its successor, *The Story of Esther Costello*, being filmed, and a new bestseller, *The Tribe That Lost Its Head*, soon to appear, he should have been at the peak of his writing power, anchored at his desk in the great stone mansion in Quebec Province that success had allowed him to buy. But, at that moment, domestic turmoil made escape imperative. He knew he had to get away, a long way away—preferably by ship.

As a childhood sailor and wartime Naval officer, he still felt that the sea was the cleanest if not the kindest of elements: the washeraway of earthbound miseries.

Miraculously, after five years of living up to a lavish reputation, he found he had cash enough in hand to sail a very long way indeed—around the entire globe. The principles instilled by a solidly middle-class Liverpool upbringing and the rectitude of Winchester schooldays demanded, however, that there should be what he called a frugal excuse for such extravagant self-indulgence. And so, gradually, the idea emerged of visiting all the chief harbours of the world and using them later in a book.

His publishers were eager for another sea story. He, not wanting to be typed as a sea-writer, had been determined to go on exploring other themes but, as his odyssey from Norway through India and the Far East to Australia and on to San Francisco, the Great Lakes and Newfoundland progressed, the ultimate maritime celebration began to take hold of his imagination: the centuries-long story of all that sailors had done to discover, open up and police the world, four hundred years of history, from Drake to the present day, written as a novel.

It was a project so massive that it took twenty years to mature, and was sometimes to scare him with its scope and the research needed even to begin.

While writing other books at the rate of one a year, he stocked his library with rare leather-bound volumes on the circumnavigation of the globe and the slang used by sailors through the ages. He read, sorted, planned, made what most authors would consider absurdly few notes, and digested a great mass of detail so thoroughly that it became more alive to him than his own day-to-day experiences.

It has been said that he put off writing the book for so long because he knew it would be, as it tragically proved, his last; and this is true, though not for the psychotic or visionary reasons that have been suggested.

The real cause was simpler and more professional. He knew he would never have a better idea. He wanted to do it justice, to make it the pinnacle of his career. He did not like to think of following it with anything inferior in scale.

And so he delayed—just a fraction too long. During the years after he began writing the first section in 1974 he was never completely well, though unaware of the cause or of how serious it was to become. The words came more slowly, but he followed throughout the working schedule he had established through the years: getting up soon after five each morning, thinking, listening to several bulletins of radio news, inspecting the garden, and downing a curiously liquid and widespread breakfast that began with fruit juice, progressed to milk, and ended with two lager beers.

The garden was a late-discovered delight. He enjoyed watching things grow. The miracle of a few seeds producing a crop, and a crop, by dint of savings in the weekly budget, helping to pay for the gardener, entranced him. Flowers didn't move him to anything near the same extent, except for a few of the more exotic varieties. He would have been perfectly happy to have potatoes growing up to the sitting-room windows.

Entirely by himself, through careful and at first incredulous observation, he discovered the role of bees. The ingenuity of the system delighted him and, when he heard a radio talk on the sex-life of snails, he decided he must some day write a book on the usefulness and procreation of insects and other oddly behaved species.

The whole of nature, springing on him unawares in his sixties in Gozo, where growth is satisfyingly speedy, was an enchanting revelation which led him, in *The Kappillan of Malta*, to make his priestly hero, Father Salvatore, end his days tending a monastery garden, producing more beans, onions and melons than the monks could ever eat. It was an escape from the world, when the world pressed in too hard, that he would have relished for himself.

However, at nine each morning, no matter how tempting the current crops, the serious work of the day began. Upstairs in our sun-soaked, golden stone, many-arched and court-yarded farmhouse, in an office that looked over fields—other people's fields—to the blue of the Mediterranean, he corrected for the final time the two pages, the six hundred words, he had written the day before, fair-copied them, and began work on the next two.

At 12.30 he stopped for a mid-day drink in the sun and, after lunch, he slept again, a split sleeping shift he had learned in the Navy and held to whenever he could for the rest of his life.

In the evening he had another working session after dinner, from nine until midnight or beyond, his best writing time, when the house was still, the telephone never rang and neither man, dog nor goat stirred in our tiny farming village.

Every paragraph he ever wrote was corrected ten times or more before he was satisfied enough to fair-copy it. After that it was rarely touched. On several occasions, towards the end of a book, when his publishers were waiting in restrained desperation for the last chapters, the pages were sent off virtually straight from the typewriter to the printers.

For *The Master Mariner* he set himself an additional discipline. Each section was written with the lilt and flavour of the age of which he wrote, and he checked every important word to make sure it had been in general usage within ten years of the time he was describing.

Before he began to write any of his novels, the whole book

was so immaculately planned that he never needed to go back over completed chapters to add an extra scene or explanation. That, he said, was his civil service training, an apprenticeship that also accounted for his passion for lists—lists of things to be done, lists of the number of letters received each day, lists of how many cucumbers and tomatoes were gathered from the garden in a season, lists of the number of blooms on a favourite hibiscus in a year, lists of expenditure, and an account book divided into an infinity of zealously tended headings.

It was typical of him that he referred to himself as a former civil servant rather than a diplomat. The appalling poverty uncovered by the Invergordon naval mutiny, not long after he came down from Cambridge, sparked in him a crusading belief in socialism that was to dominate his middle years and leave him forever sensitive to the nuances of rank and privilege.

He later came to know and greatly admire many diplomats, but the echo of the early scorn he had poured on stuffed shirt die-hards at street-corner party meetings and in his only play remained with him. For himself he preferred the humbler label —civil servant, spoken with a wry smile. Also, he liked the idea of service.

When it became clear, soon after it appeared, that the unstoppable sales of *The Cruel Sea* were likely to make publishing history, he had been a diplomat (or civil servant) for five years, having gone to South Africa at the end of the war to set up and then run the British Information Office in Johannesburg. He could well have afforded to give up his £2,200 a year government salary and resign, but instead he stayed for another two years in Africa and then moved on to Ottawa for a further three years, to do the same information job there.

He would have remained longer still if certain cautious members of the Commonwealth Relations Office had not shown what he considered a wounding want of trust in him by making it plain that they would need to see, and almost certainly to censor, the manuscript of *The Tribe That Lost Its Head*, the book he wrote, at an unfashionable time for such a theme, to honour the work and dedication of British civil servants abroad.

The Tribe, set in a mythical African country, was written in Ottawa, thousands of miles from the continent that had inspired it. *The Cruel Sea*, his tribute to the Royal Navy, another service of which he was proud to have been a member, had been written at an equally odd distance from its source: in Johannesburg, 6,000 feet above sea level.

He was eternally grateful to the Navy. At the outbreak of war he was a doctrinaire socialist turning pink at the edges, a pacifist and an impoverished free-lance writer, eating one solid meal a week by courtesy of the magazine for which, as a matter of survival, he was restaurant correspondent.

At twenty-three he had turned his back on the comfortable Liverpool home of his surgeon father, fled from his desk in his uncle's Nottingham solicitor's office, and taken his typewriter to a single, mildewed room in London where he became a Pimlican and an embryo novelist.

When war was declared, confident that the bombs would immediately begin to rain on London, he joined the St John's Ambulance Brigade, whose uniform was the only one his pacifism would allow him to wear. But the bombs did not fall and he soon came to the uncomfortable conclusion that once war became a fact one had to fight the enemy first and look to one's moral principles afterwards. He joined the battle by answering an advertisement in *The Times* advising 'Gentlemen with yachting experience' to apply for commissions in the Royal Naval Volunteer Reserve.

After a few weeks' training he was sent to sea on Atlantic convoys and worked his way up from Temporary Probationary Sub-Lieutenant in corvettes to Lieutenant-Commander and captain of a frigate. Almost every incident he was to record in *The Cruel Sea* he wrote from first-hand experience. But still he thanked the Navy. It had, he said, taken on a scruffy journalist with three bad novels and one good one to his credit, and made a man of him, returning him, as he wrote in his autobiography, to the early-learned virtues and constraints that for more than a decade he had consigned to the attic of derision.

The three bad novels had earned him no more than £30 a piece. The one good one, *This is the Schoolroom*, which told all he had learned by the age of twenty-nine, had been published three days before the outbreak of war—not a good time to attract the public's attention.

His play, *The Visitor*, staged in London in 1936, with Greer Garson in the lead, had survived for only three weeks. Soon afterwards, the theatre, Daly's, had been pulled down and replaced by a cinema, and Greer Garson had left for Hollywood, never to return. *The Visitor*, wrote Nicholas later, had enshrined some of his then dearest convictions and been based on a profound misconception of human nature. Sadly, no copy has survived.

When Nicholas first conceived *The Master Mariner*, he imagined it as one volume encompassing the whole sweep of maritime development from the time of the Armada in 1588 to the opening in 1960 of the St Lawrence Seaway, the farthest penetration of the land (2,000 miles) ever made by ocean-going sailors. But it soon became obvious that such a book would, through sheer weight of pages, be unpickupable rather than unputdownable. And so, reluctantly, he decided to split it into two.

There was also the problem of how to link so many disparate times and characters. He discarded the obvious one of a seafaring family saga. There would have had to be too much genealogy, too many purely fictional triumphs and disasters. The finished book was to be The Sea, he said, not a floating Debrett. So, eventually, he devised a combination of the Wandering Jew and the Flying Dutchman: a young Devon sailor, Matthew Lawe, cursed after a spectacular act of cowardice to wander the wild waters till all the seas ran dry.

'Lawe is one man,' he wrote in an early résumé of the book, 'one character; he is also the spirit of maritime exploration, adventure, and fortitude, a spirit which has conquered the seven oceans of the world during the past four hundred years. His life is the thread stringing together a long history of sea-going, from Spanish galleon to modern man-of-war, from sail to steam, from wooden walls to iron-clad; and, on another plane of gallantry, from West India quay-side trollops to the fashionably frail beauties of Regency London.

'He is the kind of man, repeated endlessly down the centuries, who fought and used the sea to discover and tame the land.'

The first volume, *Running Proud*, appeared in September 1978. It told the story from the time Sir Francis Drake left his game of bowls to fight the great crescent of the Spanish Fleet to the death of Nelson at the Battle of Trafalgar two hundred years later, taking in on the way Henry Hudson's search for the North-West Passage, cut-throat piracy in West Indian seas, Captain Cook's contribution to Wolfe's victory on the Heights of Abraham and his own great voyages of discovery, cod-fishing off Newfoundland, and, as a landbased interlude, lascivious Samuel Pepys's achievements in Naval expansion and administrative reform.

It was his greatest success since *The Cruel Sea*, which was, he said, a very nice thing to happen at the age of sixty-eight. *Running Proud* reached the top of the best-seller lists ten days after publication and stayed there for months. *The Times* called it 'a rich, rare and noble feast'.

The second volume, *Darken Ship*, was to have shown Matthew Lawe, still an accursed coward, still ageing at the rate of only ten years each century, continuing a pattern of fluctuating fortune: growing rich and shamed as the captain of a slave ship on the infamous round trip between Liverpool, the Slave Coast and Barbados; reduced to the ranks again as pressed man aboard the *Shannon* frigate just before her famous duel with the American *Chesapeake;* speeding under an elegant spread of canvas on a clipper ship from Australia, this time with Conrad as First Mate and Galsworthy a passenger; searching for the remains of Sir John Franklin in the Arctic; fighting for his country in wartime, and suffering, in peace, the sailor's eternal fate of neglect.

Before he died last year Nicholas had written only the first section of *Darken Ship*, in which Matthew Lawe rises to unaccustomed prosperity through his dealings in human flesh. Though it would have formed only a seventh of the finished book, it is immensely rich and varied, with splendid sea episodes, the horrors of an inhuman and recently illegal trade, and the most accommodating and tireless of female enchantresses.

Even this chapter is not quite complete. The finished typescript stops on a high peak of drama: Matthew Lawe boarding an early nineteenth-century *Mary Celeste*, a ship manned only by corpses. He is about to enter the captain's cabin to seek the reason for such a gross swath of mortality . . . and there the typescript ends.

When he was working well, Nicholas gave me his finished pages to read every few days in batches of ten or so, but towards the end I think he didn't like to admit how little he was producing. I hadn't seen anything for several weeks and when I read that last scene in the folder he had had with him in his London hospital I thought it was the end and that there never would be a satisfactory explanation.

When I returned to Gozo, however, on the lectern in his office, which he stood at to plan and polish each draft, I found two more foolscap sheets, typed in red, like all his unfinished work, and corrected by hand. These provided the answer.

There were also, of course, his notes, giving more details that would have been woven into his last few remaining pages and, most important of all, outlines of this and each of the other six sections that would have made up the book.

These synopses are unusually full and carefully written since they were prepared for a television company interested in producing a series based on *The Master Mariner*. But for that there would be little to show how Matthew Lawe's story was to unfold, for, though Nicholas planned and plotted so carefully, by far the greater part of this preparatory effort remained as a rule filed away in his brain. Only the briefest of headings and the most clinical of notes were usually committed to paper.

All these various strands are now published for the first time in this volume. The nearly finished section is rounded off with the last few pages of its synopsis, and the outlines of all the other chapters, encapsulated history, are included, bringing to an end the story of Matthew Lawe's wanderings.

There is only one major change Nicholas might have made

and that is to the ending of the entire saga. When he devised the final chapter its setting, the St Lawrence Seaway, was newly opened and we were living on it, on one of the Thousand Islands, a small rock with a house on top and great tankers from twenty nations sweeping past our front door. That daily and dearly loved view, together with the flowering of the folk-song era, inspired the idea of a young guitarist sitting beside our five-mile-wide river, singing 'Till all the seas gang dry', an echo of the Scottish witch's curse that had condemned Matthew Lawe to his four hundred years of adventure, fear, exploration, love, loneliness, modest wealth and wretched penury.

Nicholas might well have followed through and written it that way, but, in Malta, where he lived for the last ten years of his life, he renewed again his acquaintance with the Royal Navy and learned to know well many a visiting warship and a few of the mysteries of modern weaponry.

He was already toying with the possibilities of these when a seafaring reader wrote commanding him to justify a phrase from some early publicity material describing what was to be included in the sequel to *Running Proud*, words which could have been construed as meaning that in the present day the sea is no longer a violent element to be conquered.

Nicholas, of course, did not write his own advertisements, but he did answer every single letter he received, and so he replied:

'I agree that the sentence you query would have been better phrased as "*not only* a violent element to be conquered, but *also*, host to the most destructive inventions of man", since the particular violence of the sea never alters.

'But what we cannot yet gauge is the future contrast between these two dangers. You and I know the condition to be met in the Pentland Firth or off Sable Island. What we do not yet know is the effect of, for example, an atomic bomb which might vaporize the sea over an area of ten miles by ten miles by five miles deep, with a convoy of a hundred ships sliding into the middle of the pit. That is what was meant by the destructive hazards of the future.'

In such a way as this the seas might well have run dry enough to release Matthew from his purgatory.

There is one more small item shown in facsimile at the beginning of this book, a note jotted down at some time during the last few years, and tucked haphazardly into the slim, blue loose-leaf folder that contained all the notes for *The Master Mariner*. Nicholas intended to use it in his final chapter, when a weary Matthew was to dream at last of peace, an end to his years of struggle, a short space to indulge in the innocent delights of beachcombing. This was, of course, Nicholas's own dream: to be the man at the edge of the tide-mark—the phrase with which he began *Running Proud*—with no burdens, no debts, no six hundred words to be produced that day. He was very nearly in a position to achieve it.

But I don't believe, had he lived, that he would have been content for long to let *The Master Mariner* remain his last book. He already had plans for a third African novel—this time showing an African country winning through to stability and achievement—and a newer idea was also taking shape: the story of Pontius Pilate, whom he saw as a civil servant promoted beyond his capacity, knowing that whatever he did with that particular problem was going to be wrong. The story of a civil servant, written by a civil servant.

Even after *The Master Mariner*, he would have continued to fight against being cast exclusively as a sea-writer, but the sea, without doubt, would have continued to lure him. On his early-morning beachcombing expeditions, the tide one day would have washed up yet another maritime figment that would so have lodged itself in his brain that he would have been driven again to his typewriter, if only to free himself of it. As an admiring reader once told him: when he wrote of anything he wrote well; when he wrote of the sea an extra magic took over.

Ann Monsarrat
Gozo, January 1980

eight

Darken Ship
1808

'A man of the name of Nowell, who lives in St Andrew's Parish, Barbadoes, had been in the habit of behaving most brutally towards his wife, and one day went so far as to lock her up in her room, and confine her in chains. A Negro woman belonging to this man, touched with compassion for her unfortunate mistress, undertook privately to release her.

'Nowell, on discovering the poor creature had been instrumental to his wife's escape, obliged her to put her tongue through a hole in the board, to which he fastened it on the opposite side with a fork, and afterwards drew out her tongue by the roots, of which she instantly died.'

The Horrors of the Negro Slavery, from Official Documents recently presented to the House of Commons.
Printed for J. Hatchard, Piccadilly, 1805

'It was the unique achievement of Britain to lead in the complete elimination of the Trade. Control of more than half its total volume marked her out for a special contribution; sea-power gave her power to enforce abolition; and religious enthusiasm inspired her citizens with crusading zeal . . . In 1807 Wilberforce carried a bill in Parliament, which forbade British subjects and British ships from taking any part in the slave trade.'

Chambers's Encyclopaedia, Volume XII
under 'Slavery and Serfdom'

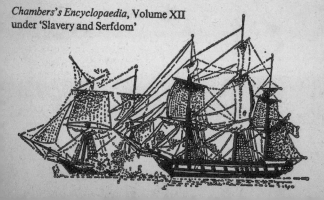

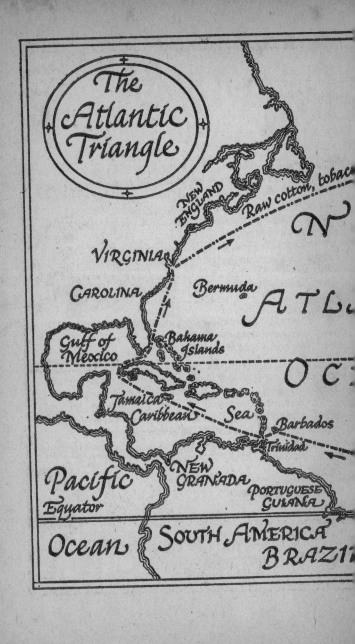

The Atlantic Triangle

NEW ENGLAND

VIRGINIA

CAROLINA

Gulf of Mexico

Bahama Islands

Bermuda

Raw cotton, tobacco

N

ATL

O

Jamaica

Caribbean

Sea

Barbados

Trinidad

Pacific

Equator

NEW GRANADA

PORTUGUESE GUIANA

Ocean

SOUTH AMERICA

BRAZIL

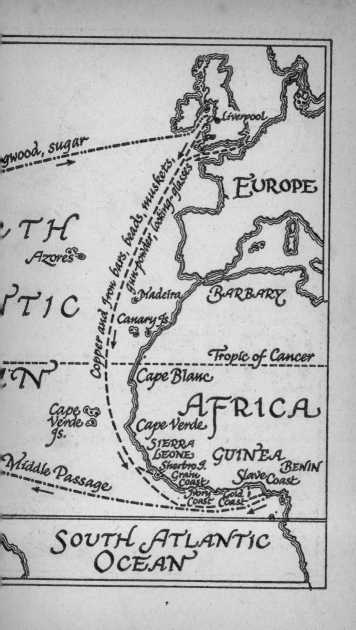

gwood, sugar

Liverpool

EUROPE

Azores

Copper and iron bars, beads, muskets, gun-powder, looking-glasses,

Madeira

BARBARY

Canary Js.

Tropic of Cancer

Cape Blanc

AFRICA

Cape Verde Js.

Cape Verde

SIERRA LEONE

GUINEA

BENIN

Sherbro J.

Slave Coast

Grain Coast

Middle Passage

Ivory Coast

Gold Coast

SOUTH ATLANTIC OCEAN

1

The bitter cold morning of the ninth day of January, 1806, signalled a most curious division between great and small in much of southern England, and also gave ample evidence of a shared bereavement. Though the city was full of common sailors, come up in their thousands from Plymouth and Portsmouth and a dozen other seaports to witness the burial of the nation's beloved hero, they were scarcely to be seen as Lord Nelson's funeral car passed in procession through the streets of London.

There, all was pomp and ceremony and the mourning display of great men, of whom some had grown in his shadow, some had loved him, some had censured and sneered—and lived to bury him. While sailors wept on the outskirts of the crowd, thirty-one admirals, one hundred sea-captains, lords without number, ladies without credit or repute, heralds magnificent, bishops a-blessing, rode by in gilded coach and emblazoned carriage from Whitehall Stairs to St Paul's Cathedral, to give honour, or to take it, on this most solemn day.

Along with them marched ten thousand soldiers. *Soldiers?* Yet they were all mourners, whether they shed the tears of the crocodile or the heart's blood of desolation. Small Nelson in his noble coffin bound their multitude together. But it could be seen and felt, that for all the finery and show the crowds warmed most to the sight of forty-eight men of the *Victory*, the Admiral's own humble shipmates, plodding manfully along at the tail of the procession, bearing the shot-torn colours of his flagship.

The sailors' gait was unmistakable—and, within the heart the most moving—in all the throng which had on that day the privilege of St Paul's and the pride of witness.

Thus was his Lordship laid to rest, to the mourning of England and the orphanage of sailors. At this last moment, all his accounts were settled, whether good or bad, gleaming-gold or gaudy-scandalous. There remained of his life a wonderful balance in hand, which would (it was said) serve British sea-power, and with it the lawful fabric of the world, for a hundred years or more.

Yet simple people, stumbling away from sorrow, could only feel within their barren breasts: 'We would rather have lost Trafalgar, and kept the man.'

One such barren-breasted mourner was Matthew Lawe: naval lieutenant, branded coward, doomed wanderer of the sea, now surviving more than two hundred and forty years, yet with the body of a thirty-year-old. It was not to be believed, save within that enduring body, and the heart which, beyond any doubt, had driven it onwards since the days of the Great Armada.

One need do no more than suspend belief, as this luckless culprit had been forced to do; to live on, to suffer, to rise and fall like the endless tides: and, on this despairing funeral night, to watch from the shadows and then quit the scene, seeking the only bearable solace, the company of other sailors.

Fortune favoured him. There was a tavern in the town, below Thames Street on the water-front, and there, between despair and drunkenness, the tears of sorrow and the rage of guilt, he had luck enough to find the men he sought.

They were a company of sailors, keeping the mournful hours of the middle watch. On this extraordinary night, they wanted very little: *not* women, *not* landsmen, *not* anyone who did not speak their language. They wanted liquor, but not even much of that. They wanted to remember, and then decently to forget. They wanted Nelson, and they would never have him again.

Matthew Lawe, out of uniform and into the sober black of sorrow which was the world's wear on that day and night, came hesitantly through the tavern door, to find immediate comradeship, warm and comforting as an arm round the shoulder. He was hailed—and he had never been more moved at any salutation—by a group of men who sat apart: pig-tailed men, men in shabby blue, horny-handed men with faces creased and weather-worn: his very own kind, and thank the merciful Lord for that!

'Mr Lawe!'
'Come join us!'
'Make a space there!'
'How does your honour?'

'By God, lads, we are all shipmates again! '

Matthew could have wept, and for a brief space he did so, and at this forlorn hour it was nothing new. Tears were the rig-of-the-day, and the only men to be ashamed were those who did not shed them. But confused by liquor, and with blurred eyes, he found that he did not instantly know his company.

'Are you *Victories*?' he asked, when he had sat down, and sipped at a mug of ale, and smiled at such loved company, and looked from face to face. 'You must bear with me. I am not in the best shape tonight.'

'Who is?'

'Who could be?'

Then he found that, through the smoke and the gloom of the tavern, he was staring directly into the eyes of a man he recognized, from twelve years earlier, from Lord Nelson's first ship-of-the-line, his first great command, his favourite because she had been so hardly won after years of useless life ashore. The man was square, and strong, and solid as the oak of that ship. Such a man of authority, drinking on this night of forgiveness and friendship with the top-men and gunners whom he daily tyrannized, could only be——

'Blantyre!' Matthew Lawe exclaimed. 'Blantyre the boat-swain! So you must be——'

'Old *Agamemnons*, Mr Lawe. All of us. And blessed glad to see you again!' He touched his forehead in salute, and held out his hand, which Matthew almost beyond words or manly feelings willingly took. 'We clubbed together, and came up by coach from Portsmouth. We would have crawled, if need be . . . You know we are Captain Berry's men now, but then we were his Lordship's.'

'And ever will be!'

The man who spoke so eagerly was also to be recognized: one John Pasco, a gun-captain, a Cornishman, though dark with the blood of those long-ago Armada castaways. He also held out an iron hand and gripped Matthew's as if it had been a ramrod stuck in the muzzle. Then he asked: 'Sir, do you mind when you first spoke to me?'

'Aye.' It was easier now: the memories came flooding back

to Matthew as if they were yesterday's. 'That famous ripple-firing. The first practice broadside.'

'I thought it would be my last . . . I never saw his Lordship in such a passion. I heard him on the quarter-deck afterwards, when he said: "We will fire again. *And again!* " He would have fired *me* if he could have stuffed me into a cartridge!'

'But what happened?' a voice from the shadows asked.

'Something went amiss,' Pasco answered brusquely. Perhaps the man who questioned him was not worthy to share a wounding confidence. 'The flint broke and fell, before it sparked. Instead of a shot, there was a silence. One could hear it all over the ship. Then Mr Lawe said: "Fire number three, damn your eyes!" and by God three cannon went off at the same time! I could have died on the spot.'

At the mention of death, Lawe said quietly: 'I may assure you, all is forgiven tonight.'

'Aye,' said Boatswain Blantyre. 'But by God, what a price!'

The price was still with them all.

A pause for drinking, and then the eager questioning began. They recognized in Matthew Lawe a man who could ease their need to know. Gun-deck sailors saw little of the larger battle, even from their own ship; they spent their fateful day chained to one weapon, one gaping gun-port, one chance of death or wounds. Lieutenants—and especially this lieutenant, who had been so close to Lord Nelson—could tell them more, tell them all, make the shadowy picture clearer, and perhaps more bearable.

Blantyre said: 'Sir, you must have been with him when he fell.'

'As near as I am to you. Mr Scott his secretary had been killed at his side, and thrown overboard. I was in the cockpit with the wounded, and sent for to take his place.'

They waited avidly. But now it was more difficult. Truth must be guarded, and folly and cowardice masked, while leaving a picture good enough for these seasoned men, and for their grandchildren, and the kindly frame of history. As he hesitated, gun-captain Pasco asked:

'But how was he shot?'

'From above. A musket-ball through the shoulders, and down through the breast to his backbone. He fell at once.'

Matthew was wincing with the pain of it, and so perhaps were his hearers. Blantyre asked:

'So you were by his side?'

'Aye. Not a yard away.'

It was the first lie.

'Did you see the man that shot him?'

'No.'

This was his second lie, and the most shameful. He had indeed seen the marksman, a French marine in the mizzen-top of the *Redoutable*, taking careful aim at a splendidly garbed Admiral—or at himself. Within a moment, without a word of warning, he had run for cover beyond the mainmast, and from there turned to watch the abandoned sacrifice—Lord Nelson falling stricken on his quarter-deck.

But how could he tell such a tale to these men? How could he tell St Peter, or God, or Christ the Redeemer? Neither man nor God would do less than spew him out forever.

One little man, at least, seemed prepared to do so already. There came a voice from the smoky darkness, the same somewhat saucy voice as had asked Pasco the gunner what had happened at the ripple-firing. This time it said:

'If we had kept him aboard the *Agamemnon*, he would still be alive! '

Matthew stiffened, for a moment cold with anger rather than guilt. The charge might be true, but it could not be allowed to pass, and bring public shame on the *Victory*. He leaned forward, staring into the blank from which the voice had come, and demanded sternly:

'Who are you, lad, that makes so bold? Show yourself!'

In the hush that followed, a face came into sight at the far end of the tavern table. It was the kind of face which Matthew Lawe had expected: a young face, mean, mocking, flushed with liquor. But its owner had lost some of his assurance as he answered:

'Tom Taylor, sir. Top-man . . . I meant no harm.'

'And I mean no harm to you, Tom Taylor,' Matthew said.

'But keep your tongue off the *Victory*. We did our best. No ship's company in the world could have stopped that musket-ball. And no top-man either . . .'

Lies, all lies . . . But there was a murmur of assent from those around him: though the *Agamemnons* might have proper pride in their ship, it would not do, on this night, to insult another of the same Fleet. Blantyre, seeking to ease a moment which might tarnish their solemn meeting, said:

'Perhaps he was marked for death at Trafalgar, and there's the end of it. God knows he had cheated it, times enough . . . Did you sight the funeral car, Mr Lawe?'

'Aye. Very fine.'

'And the great gilded coffin atop of it. I never saw such a monument!'

'Very fine,' Matthew repeated. Then he added: 'But I can tell you more of what was inside it.'

'Why, Lord Nelson, surely?'

'Aye. But first there was another coffin, a little one. Made by the *Swiftsure*'s carpenter, and given to his Lordship by Captain Hallowell. Made from the mainmast of the *Orient*, French flagship at the Battle of the Nile. You must have heard how she blew up, in all that fury.'

'But the Nile was seven years back,' gunner Pasco said, astonished. 'Did he keep it all that time?'

'He kept it in his cabin! He loved it, and brought people to see it. He once said: "This is where I shall sleep—and soon." I can tell you, while that grand catafalque was for show, the little wooden one was for honour!'

But suddenly this talk of sleep and death choked him, and he could say no more, nor stay no more. The memories were too stark, too pitiful. He could see Lord Nelson's body as it had been laid in that wooden coffin: preserved more than two months in brandy and spirits of wine, tiny, shrivelled, a dark skeleton with all the glory gone, all the manhood surrendered. He could see, even more clearly, his own part in this robbery of life.

He stood up, wiped his eyes, muttered 'Good-bye, lads,' and made for the tavern door. After uncounted years, he still did not deserve the company of sailors . . . With another traitor, St Peter himself, by his side, he went out, and wept bitterly.

2

That had been some two years ago; and now, far away from London, he was waiting to keep a shameful appointment, one of his own despairing choice which matched his degradation.

Half-way up the slopes of Mount Pleasant in Liverpool there stood a new-built house, suitable to a shipowner who had prospered beyond his mother's fondest expectations. Matthew had called upon Mr Benjamin Boothroyd once before, a few days earlier, and now he was making his return, to learn if his request for employment would be granted: if he was thought worthy of the honour of command—of a new Liverpool slaveship.

He knew that he was.

Many things had forced him to this decision: many strands of thought and feeling, many wounds of the spirit, much weariness, much ambition brought low and swallowed up in failure. As to his main doom, he had ceased to care; if it were truth, then he could not fight it, if it were some foolish fevered dream, then he was still trapped within it.

The time to cope with the stress of life was *now*, on every morning of every new day; and on a certain morning some months ago, he had plotted his course as best he could.

Sailors by the thousand were out of work, as was usual after battle and victory, and were left to beg their bread on disdainful streets where their service was forgotten. But Matthew Lawe wanted no more naval duty. He did not deserve it, and could never match it. After such a miserable display as had attended the death of Lord Nelson, to whom he had owed loyalty beyond any limit, he could only fall, or bring himself down for very shame.

If fall he must (he had decided, after a night of desperate drunken wandering along the darkest alley-ways of London) then it should be into the pit. Thus he had chosen the flesh trade, that stain upon the sea which had now been placed beyond the law of the land, but which still flourished like the Tree of Evil.

31

There he would plunge downwards again—and God damn the world of honour!

The trade in black flesh still centred upon the Round Trip, that great triangle of infamy compounded of pain, perjured faith, and profit. It was a British enterprise. The ships sailed from Liverpool, and Bristol, and London, and the River Clyde, all pointing southwards to the Slave Coast of West Africa. From there they made for Virginia, or for West Indian ports of which Barbados was the principal haven; then sped back again to their homeland.

Matthew Lawe, like any ardent schoolboy, had already done his sums and, by diligent inquiry, had made himself expert in this endeavour—and envious. The profitable link in an evil chain was the Middle Passage, the slave-run from east to west. The rest was plain sailing, with a sharp look-out for ships of an inquisitive British Navy, which now had its stern mandate to clew-up the whole business.

England had been vastly enriched by this trade: Liverpool and Bristol especially had risen from fishing-ports to towns, from towns to cities, and grown fat on its profits: individual men had watched their plunder outstrip even their avarice.

Starting with a cargo of 'Manchester Goods' and such-like trash, the meanest barter-currency of all, for which a black man was readily betrayed by his fellows—anything from cotton skirts to tin kettles, brass nails to iron cooking-pots, ancient flintlock muskets to bead necklaces, sandals to straw hats, looking-glasses to fish-hooks and pellets of lead—starting with this Judas coin, and the gold and gunpowder which was the true price to the slave-broker, such a voyage could only prosper.

Once on the Coast this first cargo was exchanged for all manner of slaves, rounded up in forest barracoons by their own tribal chiefs or by Arab slave-raiders; and from there, well shackled and battened down, these were carried westwards into servitude.

Here they were once more bartered for all the riches of a sunshine island—sugar, rum, tobacco, syrup, coconut oil, molasses, and ginger-root. Triumphant captains then raced home with a profit—even allowing for natural wastage, which was human death—of any sum upwards of forty *per centum* on every voyage.

Sailing outwards with the North-East Trades, and homeward again with the Gulf Stream and the Atlantic westerlies, such a Round Trip might take a year or more. Sailors' talk proclaimed the most important man on board, after the Captain, to be the ship's surgeon who, with luck, skill, and sobriety might outwit all the hazards—disease, brutal punishment, seasickness, hopeless despair—and deliver to the Barbados a live slave instead of a worn and empty shackle.

He might also deliver a chance commodity to be later noted in the manifest as 'With Child at Breast'.

At £45 a head for a fine black buck in prime condition, the difference between life and death was a measurable blessing—or, as pious shipowners put it, with no trace of irony, a God-Send. A brand-new slave-ship might pay for herself in three voyages. A gain of £60,000 on one single Round Trip had already passed into the history of commerce.

Naturally there were always hazards, and now these had multiplied. A year earlier, this cannibal trade had been abolished in England, and any man who engaged in it did so at his peril. His ship might be chased and boarded by the Royal Navy, the slaves set free (with total loss of cargo), the vessel itself confiscated, and the crew press-ganged into more reputable employment.

All officers, and ultimately the owners, might suffer further penalties under the law.

England, for so long the chief culprit in a cruel game which had seen some three million slaves transported westwards in the last century alone—England had now taken the lead in this crusade of mercy, sparked by the fiery parliamentary speeches of Mr William Wilberforce and the determined preaching of the Quakers. At whatever cost, in profits or popular esteem, she was beginning to sift this poison from the sea.

But her method was not Late Christian, so much as Early Commercial. The watchword now was: 'After Piety, the

Pocket!' It was clear that the first and best method of attack upon the slave trade was to make it unprofitable.

Yet the profits remained, for men determined, men ruthless, men set in their ways. Ben Boothroyd, a slaver-captain turned shipowner, was one such man; and Matthew Lawe was resolute to become another.

It was raining on the slopes of Mount Pleasant, and upon all Liverpool, as Matthew passed through a garden of mournfully dripping trees, bushes of rhododendron and laburnum, pathways of cinder grit, and into the portals of the Boothroyd house. Dusk had fallen on a cold foggy evening, with little light to soften the gloom or to bring hope of a better day tomorrow.

But once inside he was shown into a handsome parlour, with the lamplight and the log-fire glow shining on heavy polished mahogany and the sheen of velvet curtains; and here he found a welcome so warm and hearty that he knew, by instinct, that the berth on which he had staked his future was his, and that the terms of it would be struck before he left the house again.

Mr Benjamin Boothroyd, the sailor who had risen in the world, was portly and prosperous—and not to be crossed by anyone. The ship's captain who now owned his own fleet, and had settled for a softer life ashore, was still as hard as nails; a no-nonsense man with the harsh nasal voice of Liverpool, the habit of command, and no wish to share his good fortune with anyone unless that privilege was earned by a heaping measure of industry and service.

'Come in, come in!' he said immediately, as the maidservant ushered Matthew through the door. Boothroyd advanced to shake hands, then turned back towards the fireplace and the figure settled beside it. 'My dear, may I present Captain Lawe, of whom you have heard already.'

Benjamin Boothroyd was in the bosom of his family, and a right pretty bosom it was. Matthew scarcely had time to take in the encouraging words 'Captain Lawe' before he was bowing to Mrs Boothroyd, small and bright and plump, and returning with all solemnity the greeting of a quartet of young children

—two girls, two boys—who with a new-found gentility stood up in a row, bowed and curtised, then sat down and stared fixedly at the newcomer as if he had a long-tailed parakeet on his shoulder.

They were all dressed as small images of their parents: the boys in tight white breeches, silver-buckled shoes, and little blue tailcoats, the girls affecting the French style of clinging skirts, high bosoms—or rather, a high loading-line where the bosoms would one day sprout—and twinkling satin ballet slippers.

It was clear that no expense was being spared to fit them out for the voyage of life, of which tonight was a small beginning.

'Now, children,' Mr Boothroyd said, as if he were instructing a class of midshipmen, 'remember your manners! Harry—set a chair for the Captain. George, a glass of wine—and do not spill a drop. Effie—bring the seed-cake, and offer it. Dulcie—the dish of caraway biscuits, if you please.'

It was all done in brisk order, with bright smiles and a great display of energy. When Matthew was installed, with his hands laden, his knees balancing a plate, and his right ankle discouraging a possible enemy—a grim-faced spaniel sniffing at his boots—Mrs Boothroyd embarked on the polite conversation of a hostess: a role perhaps new to her, but already mastered in all its bottomless fluency.

First there was the weather: Liverpool weather, notorious for its horrid character. Winter lasted so long. One could only hope for spring, *which must come*. Did he not agree? But did Captain Lawe suffer from colds? She herself . . . And of course the children . . . Little George had gone down with a dreadful head cold . . . Little Harry the same, but more on the chest. Effie and Dulcie were tormented by chilblains . . . Even the servants snuffled abominably . . . Servants were so difficult to come by—and so ungrateful . . . They expected coal fires burning away at all hours . . . Did Captain Lawe know the price of coal? Indeed, the price of *everything* . . . Farm eggs were three pence a dozen . . . Butter—butter was scarce worth the buying, with Cook spreading it as thick as cheese even for the servants' suppers . . . The tradespeople had grown so grasping. Even impertinent! There had been a shopwalker at Evans the haber-dasher, only yesterday, who had gone so far as to . . . But she had given him a piece of her mind, make no mistake! . . . Would Captain Lawe take another glass of wine? It was a special consignment from Madeira. Mr Boothroyd had so many friends in the wine trade. Of course, one did not have to depend on such favours . . .

In a rare pause among this enthralling chatter, during which Matthew had time to reflect: how sad that a pretty woman must gush like a garden tap left open, the youngest child, Dulcie, made bold to interrupt, though with extreme caution.

'Mama,' she breathed hoarsely, 'may I have a biscuit?'

Benjamin Boothroyd, who had been silent, watching Mat-thew, came to attention. 'You may each have *one* biscuit, or *one* thin slice of seed-cake. Then it is time for bed.'

The children, released, crowded round the sideboard, with much whispering and elbowing. From there Harry, the elder boy, turned to say:

'Papa, may I ask Captain Lawe a question?'

'If it is a sensible one, yes.'

Young Harry manfully believed it to be a sensible one. 'Sir, is it true that you fought at the Battle of Trafalgar?'

'Why, yes,' Matthew answered, grateful for a change of style. He was sweating from the heat of the fire, his glass was again empty, he had crumbs on his trouser knees, and the spaniel, satisfied with his boots, was giving certain warnings of incontinence. 'I had that honour.'

'Was it very fierce, sir?'

'The fiercest of my life.'

'Were you very brave?'

'Now Harry,' his father interrupted, 'a foolish question only deserves a foolish answer. If Captain Lawe was brave, of which I have no doubt, he would not say so, would he?' He stood up, a comfortable, confident figure by his own fireside. 'Children —off with you to bed! But first, say goodnight to Captain Lawe.'

The little line-of-battle was re-formed, the well-schooled bows and curtsies repeated. When they were gone, Matthew said:

'You must be proud of them, ma'am. I have never seen such perfect good behaviour.'

Mrs Boothroyd preened herself. 'Too kind, I am sure. But yes, they do well enough . . . Of course, they attend a school of deportment, with all their little friends. I think they can only profit by it . . . In these days of bad behaviour among the labouring classes . . . It is important that families such as ours set a tone . . .'

They supped plainly and heartily off brown soup, game pie, 'our own garden kale', and a round of red Cheshire cheese. After that it was time for household prayers: the servants filed in, sat still as mice while the master of the house intoned the Gospel, turned to kneel against their chair-seats with their rumps boldly presented as the Lord's Prayer was repeated, and filed out again after a dutiful chorus of 'Good night, sir—good night, ma'am.'

Then Benjamin Boothroyd and Matthew Lawe retreated to an oak-lined smoking-room and the conversation, like Boothroyd himself, grew more forcible.

The change from *paterfamilias* to sharp-talking man of business came without warning: not even a wooden wand was waved as Boothroyd began:

'I like your style, Lawe. But style's not enough to go by. I've made plenty of inquiries about you, never fear. They all tell me the same thing: that you are a sailor through and through, and know how a ship should be worked. So you're the man for me.'

He made this last declaration with a great air of consequence, as if it were Matthew's passport to heaven. Matthew wished that he did not warm to it so readily . . . But he could only be glad of employment, even of the promised quality. He said:

'I must thank you, Mr Boothroyd. It is good news indeed. I will do my best to deserve it.'

'No man can do more . . . But there is one question in my mind.' Now he was looking squarely at Matthew, with searching eyes. 'You have been a Navy man, and there is no dishonour in that. But now the game—*my* game, and soon to be *your* game—is changed. Now the Navy is against us. A most disgraceful turn! I wonder what your Lord Nelson would say,

37

to see the service sunk so low as to interfere with our lawful commerce. The trade in slaves is no shame to anyone. It has been the foundation of this city! Now they would steal it from us. By Jupiter, they will be making slaves of *us* next!'

Matthew had met this outraged talk before, all over Liverpool, from those involved in the trade and others dependent on it. The Bill of Abolition had struck such merchants very hard. Though some were now content to obey the law, under protest, and to search for their profits elsewhere, others swore that they would see all snivelling saints hanged, before they submitted to such tyranny. Ben Boothroyd was one of these, and he wished to be sure of his ally. Matthew answered:

'If I take command, I will take all that proceeds from it.'

'I want no Quaker nonsense in my affairs,' Boothroyd said harshly, 'and I want no Navy nonsense either. You are to forget the Navy, and you serve *me*. Agreed?'

'Aye.'

'Well enough. I'll say no more on that foul subject, since I take you to be a man of honour.' The fateful word hung in the air between them, while something within Matthew's breast shrank smaller still. *His* Lord Nelson had wagered the wreck of his career—had spent five years ashore, beached and dry— for trying to enforce such trading laws in the West Indies, when the old Navigation Act had barred the American colonists from their 'lawful commerce'. Honour had turned foul indeed . . . 'Now to business. I have a fine new ship, the brig *Mount Pleasant,* named by my good wife to honour our new house. You have the command. Above full wage and board, you will have a share of the profits of each voyage—three *per centum,* which in a few years will make you a rich man. But no trading on your own account, if you please. You command the ship, and conduct her, but her bottom is mine! . . . There is no need for private articles between us. My word is as good as my bond.'

He paused, while Matthew, his sad heart still in the past, remembered a tart maxim of Mr Samuel Pepys: 'If a man says to you, "My word's as good as my bond", *take his bond!*' But it was too early, and he too new to promotion, to quarrel about such terms. Mr Pepys was dead a hundred years, and all his precepts with him. Instead he said:

'The *Mount Pleasant*—is she afloat yet?'

'Afloat?' Boothroyd answered sharply. 'She is loading already, at the Salthouse Dock. Loading with suitable trade goods. You will take my meaning. I did not build her to carry nosegays to Africa—nor coals to Newcastle! All I require of you is close bargaining, small wastage of cargo, and a swift passage out and home. I fancy your ship will be fast enough to show a clean pair of heels *to anyone*. All other matters are for your own wits to solve.'

Matthew's doubts showed in his face as he answered: 'The ship I can handle. But striking bargains will be new to me.'

'I know that,' Boothroyd said curtly. 'But you will learn. And you will not be friendless.' With the stubby thumb of his right hand, and the fingers of his left, he made a count of Matthew's blessings. '*First*, you carry a trading mate on board. Penny-man by name—and a good name too! He will cast the accounts for your inspection. Be guided by him. But watch him—those pursers are all rogues! *Second*, on the Slave Coast where you are bound, the barracoons are overseen by Portuguese slave-brokers. They assemble their stock, give you your pick of it, set a price—by the head or by the hundred—and receive it from you: in gold for themselves, and in lesser goods for the runners who bring the slaves down, and the chiefs who betray them. The brokers will try to cheat you, but not monstrously—they know that you can go elsewere if you are not content.' The thumb struck down again. '*Third*, I have a man of business in Barbados, Peter Ferreira. Another Portuguese, but honest as the day. He can afford to be so! He will settle all accounts at the auction block, take the money in trust, and give you a fair cargo of rum and sugar and suchlike in return. Ferreira will load it Free on Board—all dock dues paid. All you must do is up sail and bring it home in short order. *Fourth* and last, the *Mount Pleasant* will not be alone, nor you as her captain.'

He paused at that, watching Matthew as if he might not relish Item the Fourth and last. In the silence, Matthew found himself grown more confident; by the sound of it, he would have men enough to help him in this formidable trade, and, as Boothroyd had said, he would learn . . . But 'not alone as captain'? What did that mean? He put the question, since the other man seemed to expect it.

39

'You mean, someone else is to share the command?'

'Nay, nay. I do not set two watch-dogs, where one will serve
. . . But since you are new, you will sail in company. With my
next best ship, the *Mersey Blessing,* under Captain Downie. If
you are separated, whichever is first in the Bight of Benin will
wait for the other to join.'

There seemed no harm in this companionship. 'I shall be
glad of guidance.'

'And you will profit by it. Pay all attention to Downie.
Twelve years on the Round Trip is his recommendation. He
served as my mate for five of them. A plain rough fellow. You
may take to him, or you may not. But he knows the trade.
None better! So hearken to him, and above all, see you ship well
slaved.'

One heard something new every day. 'Well slaved?'

'Aye.' Portly Benjamin Boothroyd spread his hands wide,
then slowly brought them together again. 'Stowed as tight as
may be . . . Cargo space is money. Empty space is wicked
waste.'

After a moment Matthew asked: 'Where is this *Mersey
Blessing* lying?'

'Down river from your own, about a half-mile. She is loading
also.' For the first time Ben Boothroyd grew easy of manner,
and smiled. 'If I judge you right, tomorrow you will wish to
board your own ship first. Then seek out Downie. If he is not
aboard the *Blessing,* you will find him at the Cock, across the
quay, drinking that foul Scottish brew! You will have much to
discuss. And do not forget—Downie knows the trade.' Mr
Boothroyd stood up, and held out his hand. 'Well, I'll say good
night to you, Captain Lawe. I think we shall prosper together
. . . I hope you will sail within three weeks, and we shall all be
down to see the *Mount Pleasant* go, and to wish you well.
There is only one maiden voyage!'

Captain Downie of the *Mersey Blessing* was absent from his
ship next morning, but (as Mr Benjamin Boothroyd had fore-
seen) he was to be discovered at noon in the tap room of the
Cock Inn, seated at his ease before a tall glass of a certain am-

ber liquid. He proved to be a red-haired, raw-boned Scotsman, whose gangling limbs threatened to overflow both table and chair. He was, in the manner of his countrymen, confident of himself and upstart proud of his breeding: a Scot among the English, and God's gift to both.

But he needed courage to sustain this masquerade, and he found it, as was ever the case, beneath the cork of the nearest bottle. Already, it seemed, that 'foul Scottish brew' had taken possession of him, though the pale sun was scarcely past the yard-arm.

When Matthew, after inquiries, made himself known to the lone figure in the tap-room corner, he found Captain Downie to be in sour mood.

'Lawe, is it?' Downie returned his greeting, with an early whiff of whisky and spite most evenly mixed. 'Aye, I have heard of you. Seems I have heard little else, these last few days . . . Captain Lawe of the fine new *Mount Pleasant* . . . An officer of the Navy?'

Matthew, meeting a hostile glare, replied that this had once been so.

'I've no use for them,' Downie said contemptuously. 'Great display, great swords and cocked hats, *great middens*! . . . And now spoiling our trade as if we were footpads . . . What do Navy folk know of earning their keep at sea? If they were not paid with great gobs of tax-money, they would starve! . . . Do you call yourself a sailor or a gentleman? Eh? Eh?'

Matthew was not going to have this for a single moment, cost what it might. 'I call myself nothing,' he said, with immediate anger. 'I call *you* a fool if you judge men by their rank, while knowing nothing of their quality.'

Downie stared at him, taken aback. Then: 'What quality is this?' he sneered.

'Seamanship! What else? Do you think the navy is officered by clod-hopping farmers? I have that quality, and I am assured that you have it also. So we are both sailors, and you had best remember it . . . So—do we talk, or do I tell Mr Boothroyd that I do not care for your company, and will sail alone, and be damned to you?'

Captain Downie, though still set down far beyond his expectations, had not quite done. 'Fine words! Where are the deeds to match? How would you fare, if you sailed alone?'

'Better!' Matthew said forthrightly. 'I would know for certain where I was—and it would not be at the Cock Inn, with a blowhard to foul up my course.' He stood tall above Captain Downie, ready to remain angry, ready for anything else. 'I said—do we talk?'

'Well, well, something new in the world—an Englishman with spirit!'

'No. An Englishman. And the spirit comes not from the bottle. It lies in the word.'

Downie harrumphed fiercely, glared again at this bold newcomer—and backed away from strife. With a brief gesture of welcome, he said: 'Enough, enough. Sit ye down.'

They surveyed each other for a long moment. 'Good quarrel, better friends' was a ship-board saying which Matthew, from long experience, had often found true. But it was a rough rule-of-thumb, not a surety. High words, plain speaking, sometimes cleared the air between men who *must* live at close quarters within the prison of their ship; yet sometimes it poisoned that air, to the point of murder.

Captain Downie was not his shipmate, nor, with good fortune, ever would be. But he could spoil any voyage together, if this mood of enmity persisted. Some hint of its origins came to light with his next words. After seeing Matthew bestowed with a glass and a bottle of Madeira, Downie asked:

'Have you boarded your own ship yet?'

'Aye. Very fine.'

'So she should be, by God! Brand-new from Grayson's yard —the best on the Mersey River . . . I had thought to have had her myself.'

So that was it . . . Matthew could understand the envy well enough, though he could not guess with any certainty at Boothroyd's reasons for keeping Downie where he was, and giving the new *Mount Pleasant* to a new man. He only knew his own good fortune, since he had fallen in love with this, his first command, at the very first sight.

The *Mount Pleasant* had looked a picture of shapely strength and grace as he boarded her. A two-masted brig, fresh from the Grayson & Fearon yard, with fifty years of their repute to back her building, she lay in the Salthouse Dock like a swan among crows. He had walked her decks with pride, gazed upwards at her towering masts, touched her stout bulwarks, sniffed the tarry smell of new cordage, new canvas, stood behind the wheel like any dreaming boy and imagined her running smooth before a fair Atlantic breeze.

Below was a turmoil of storing and cargo-stowage; below were the gloomy slave-decks, their ring-bolts carefully masked from prying eyes. Below was the captain's cuddy, his own snug cabin with its brass-bound chests and polished table, its lanterns hanging easy from the deck-beams, its rack of small-arms well secured, its solid privacy and comfort. From here he would rule . . .

On deck again, he had met three men who would be important to that rule: Harkness the first mate. Harkness, a little jumping fellow with a limp, seemed eager to please; Pugh looked a brawny villain—which was nothing new in his rank; and Pennyman proved a sharp-eyed, smiling rascal with a true purser's nose: pointed like a ferret's, sniffing profit from afar.

Now Matthew had Captain Downie to add to this gallery. But the ship herself was a beauty.

With some idea of mollifying the man who had failed to win her, he stretched the truth a little:

'Mr Boothroyd seemed to have good reason for keeping you in the *Mersey Blessing*. He said there was no man who could handle her better, nor ever would be.'

'Did he now?' Downie looked more suspicious than flattered. 'Ben Boothroyd is not always so free with his compliments, as you may find . . . Did he say aught of profits? Did he promise three per cent?'

'Aye.'

Downie smiled his wolfish smile. 'Did he say, "That will not stick in your gullet, will it?" '

'I fancy he did.'

'It will not stick in his, either! Three for you that does the work, ninety-seven for him that sits at home on his fat bottom, and counts his gold! By God, if I could put enough money by to build a ship of my own, I would kiss Mr Bluidy Boothroyd good-bye tomorrow!' He took a gulping swig of his raw whisky, and sat back. 'Well, enough of that—we will not grow rich by quarrelling with our bread and butter. If you are to sail in company with the *Blessing,* let's plan it fair and do it up handsome.'

He spoke fluently for the next hour, in a manner so coarse, free, and foreboding that Matthew found it hard to stomach. Was this what he would himself become? . . . It was clear that Downie was a most seasoned slaver, that his heart was in the game, for profit and a little sport besides: he talked with relish of 'black-birding', of the prudent stowage of cargo, the need for secrecy, the need for harshness, and the need for mercy—of a sort.

'Loading and unloading must be careful,' he said at one point. 'You cannot sling them up and down from a cargo hook —though you might wish to, in a certain mood. You cannot have them all whipped for keeping you awake with their moaning. Slaves can grow tender as cattle. If they stand up when landed, they will be bought. If they are cast in their stalls, they will be condemned. So see that your surgeon stays sober, and *works*! For us, dead meat is no meat at all.'

And again, as the sunlight shadows began to move across the dusty tap-room floor, and Downie called for a second bottle of each brew:

'We take on our main provisions at Kinsale in County Cork. There, 'tis cheaper, and fresher, and it lies at the furthest point south before the open sea. But all the trading cargo loads at Liverpool. You may safely sign for everything as it comes aboard. All is arranged.'

It was a matter which Matthew had already encountered, when Pennyman the trading mate, on that first boarding, had asked him to sign a whole Bible of manifests for goods already delivered.

'Pennyman is a thief,' Captain Downie declared roundly, when Matthew spoke of this. 'You should watch him as if he were a blackbird in white feathers. But he is not thieving in this matter. Here, Ben Boothroyd is the thief. He has good friends at the Custom House . . . The Manifests will all say the same thing: "General Trading Goods". You should sign them as they are writ, and no questions.'

'But without description?'

'Aye. Trading goods means slave-goods, as we know. If they are described, we might as well hoist a flag saying, "This is a slaver", and wait for the Navy to board us with cutlasses.'

'But is it not to be reported? The regulations say, all goods for export must be separately invoiced, and declared on oath, and duty paid at ten per cent.'

'The regulations say one thing,' Downie answered with the same hungry smile, 'and a good friend at Customs says another. You may believe that good friend, and you may believe Boothroyd, who spreads the grease on his palm. Boothroyd is rich, and we are not, and he did not grow rich from love of *regulations*. For us, there is only one matter to beware. Ben was a slaver captain, like us. He knows all the tricks of our trade. If we cut corners *in our own interest*, you may be sure he has cut them all before, and will meet us as we turn into the wind . . .' Downie yawned, and stretched his bony frame, and sat straight again. 'If you would prosper, get up early. But not too early. Boothroyd may never have slept!'

Once again, with a backward glance, Matthew thought: Is this what I shall become? A cheating rogue in the flesh trade? But whatever the prospect, he was committed to it, beyond question; and as the two of them said good-bye, each man to his own ship, he could at least comfort himself with a point of pride: fair trade or foul, I have my own new brig, my own *Mount Pleasant*, and her sailing will be honest, and true to the sea.

Some simple faith remained.

On a bright blustery morning in April of 1808, Mr Benjamin Boothroyd, who had the ear of the great—or a profitable way to their pockets—had assembled a notable crowd to bid farewell to the *Mount Pleasant* on her maiden voyage.

The Lord Lieutenant of Lancashire was there, and the Mayor of the city, and the Colonel of the Militia with his fine brass band, and a detachment of poor boys from the Bluecoat School, and young Mr Grayson of the shipbuilders—a little subdued, his father having been recently killed in a duel—and citizens great and small with an interest in sea-commerce or a free day to spend.

The Collector of Customs occupied an honoured place.

The band played, the brisk wind tugged at the flags and the bunting as the *Mount Pleasant*, lying alongside the dock wall but out in the tideway, poised for her departure, made ready to sail. Matthew Lawe, in smart blue pea-jacket and a peaked cap like a midshipman's adorned with the emblem of the Boothroyd Line—a chequered house-flag with the motto 'Freedom at Sea'— Matthew had little to do save smile and say farewell.

Harkness the mate gave the orders: Pugh the boatswain saw to their swift performance; Pennyman was having a last conference with certain cronies ashore. True sailors did the work.

His Honour the Mayor made a speech as windy as the weather. Then the band played 'Rule Britannia!', and Mr Boothroyd shook Matthew's hand while his four children, decked out like small theatricals, shouted 'Hip, Hooray!' and danced a little hornpipe. Then the *Mount Pleasant*, free at last, edged out into the swirling water of the River Mersey, and began her voyage downstream.

So, as captain of a ship more handsome than her line of employment deserved, at £50 a year, with all found and a share of handsome profits, an outlaw joined an outlawed trade.

3

He began this service with a memorable passage south, one of the fairest of his life.

The *Mount Pleasant* endured wild weather down the Irish Sea, took a fierce buffeting to clear Kinsale harbour after storing for her voyage, and fared still worse across the wide tormented mouth of the Bay of Biscay. But then, as soon as Matthew had beat up-wind to weather the great bulge of

Africa, and drawn level with Santa Cruz in the Canaries, his ship was free to turn due south on a gentler course; and now the ocean world was hers, and all was transformed into a magical journey into warmth and peace.

After Kinsale, they had soon lost company with the *Mersey Blessing*, which was sluggish, and no match for the new ship which Captain Downie had coveted. But why should he loiter for this laggard? . . . There came a morning when not a glimpse of her shabby topsails could be seen, across fifty miles of broad ocean, and it mattered not.

His agreement with Downie was that if they were separated, they had a rendezvous in the Bight of Benin, that harvest-home of slaves on the Slave Coast of Africa. When they were duly separated Matthew pressed on, without great guilt and with all the enjoyment in the world.

He felt full confidence, backed by the pride of command, as he romped along, alone and ahead. He could not be lost, in any sea in the world . . . He had his wits, and a ship which was proving a dependable beauty, and a good chronometer, and a gleaming new sextant bought with the last of his prize money from Trafalgar. He could sail alone for ever.

As he turned eastwards, he had also to guide him a case of sketch maps from Boothroyd's store. They showed the Ivory Coast and the Gold Coast, and the Slave Coast, all with their hills and watch-towers, distant mountains and nearer islands, reefs and rocks, landmarks which might be no more than a giant ant-hill, drawn and coloured by faithful hands of long ago. Thus he sailed on, his ship still innocent, his happiness sublime.

Coasting near the Equator, it grew hot, and strange: full of the smells and sounds of Africa, the groans of a sleeping giant, the mystery of the oldest land in the world. With time to spare —and who would count time on this timeless voyage—he overran his rendezvous, letting sea-miles slip by as far as the Bonny River and the Bight of Biafra.

By day, all was a blue and golden sky, burning heat, a dazzling spread of canvas, and an endless shrouded coast-line which stared back at him as he stared at it. By night, phosphorescent light engulfed the whole surface of the sea, so that his ship's wake spread like a fiery arrow-head behind him. Hour by hour, day by day, he was living, and loving, and learning. Little Harkness the mate, who knew these waters, added to this learning with every mile that passed.

Harkness was a Cockney Londoner, and his limping leg the product of a fall from aloft; but, as Captain Downie had once allowed, with his customary disparagement, 'he could hirple along somehow', and his skill as mate was undoubted. His was the watchful brain of ship-management, just as Boatswain Pugh was its brawny fist; together with Dunhill the second mate, they had the makings of a good after-guard.

Of the others, Matthew was less certain; Pennyman the trading mate had already shown himself a weasel, with some light-fingered manipulation of their storing accounts at Kinsale, and the *Mount Pleasant's* surgeon, one Dr Rushforth, an owner's crony and a most devout friend of anything which might come out of a bottle, had scarcely been glimpsed since they left Liverpool. If this was the speed of Dr Rushforth, Matthew thought, then God preserve him from Dr Holdback.

It mattered little, at this stage of their Round Trip, but the time for the good doctor to earn his keep was fast approaching.

How fast, Matthew grew aware as they made their passage eastwards along the shores of Africa, before turning to keep their rendezvous with Downie.

The mysterious coast had grown smoky. At night, fires blazed on the foreshore and among the trees beyond it; by day, smoke rose into the somnolent air, like successive tufts and plumes of grey feathers, as far as the eye could reach. When he asked Harkness about this, one morning watch when they were level with Accra on the Gold Coast of Ashantee, he received a ready answer:

'Sir, they tell us their trade.' He pronounced the last word as 'tryde', but Matthew had grown accustomed to this southern murder of the English tongue. 'A smoke-signal means, "Slaves are here. Come buy!" '

'But so many of them?'

Harkness shrugged his little shoulders. 'Oh, they are all at it. The charts and the pictures may say, "This is the Ivory Coast, this is the Gold Coast, this is the Slave Coast." But it is all Slave Coast hereabouts, from Cape Verde to the Congo River. A clear three thousand miles! And as soon as we show our sails, from every creek, every little river-mouth, every sand-bar, down come the slaves, driven through the rain forests and into our arms. Who sells them? *Their own gaffers!*' It was the very strangest word to hear on the coast of Africa: a good earthy English word, transplanted across all frontiers of time, climate, colour, and distance, and losing nothing on its travels. 'They would sell their own grandmothers for a little barrel of

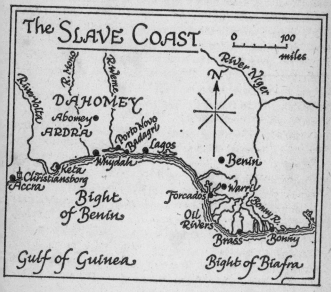

beads. Indeed, they have tried to! I can remember when a promised shipment of fifty sprouting virgins became twenty old hags too tough to be *eaten*! . . . We did not teach these blackamoors the slave trade. They have been at it five hundred years. But white man's money is what they crave. Any chief will sell his own slaves, his law-breakers, his enemies, his sons who might kill him, his wives when they are worn out, if the price suits his purse. So he is rid of them, and at a good profit.'

Harkness, after this burst of rhetoric, turned aside to speak to the helmsman, while Matthew stared at the land and wondered if what he heard was true. The port and township of Accra was left behind, and as they ran on new smoke-signals came up, tempting them into trade, proclaiming a ready treachery. It seemed that all Africa was alert for business, fair or foul, and was spreading its wares of flesh without shame or remorse. Its smoke only signalled a universal, smiling perfidy.

If it were a fact that slavery was a part of this continent, with or without the white man's help, then it was certainly a salve for the conscience.

He glanced down at the chart on the quarter-deck table, and then, under the leech of the mainsail, at the line of coast which was now shown to be 'The King of Dahomey's Land'. For the first time, they had passed east of the Greenwich meridian. When Harkness returned, he said:

'Coming up to Dahomey. Captain Downie named this for a meeting-place. You know it?'

'Aye, sir. There's a good port, Whydah, the captain's favourite. Slaves as thick as fleas, so the price is likely in our favour.'

'Is there a King of Dahomey?'

'Why, yes. And a right old villain he is! He would sell you his grandmother's *skin*, for a little something under the counter. The King of Dash, I would call him.'

It was a new word to Matthew. 'Dash? What is that?'

'A bribe,' Harkness answered. 'Everything is "dash" in Africa, and ever will be. There's a saying, you cannot haul a bucket of river water without dash for the fish! And we did not teach them *that*, either.'

Matthew was amused by his vehemence. 'What did we teach them, Mr Harkness?'

Harkness, his small brown face creased into a grin, considered briefly. But again he had his answer ready. 'To count beyond ten, sir, and thrive on it. I would not say more than that.'

They were joined now by Surgeon Rushforth, who came blinking into the sunlight from the gloomier regions below. Though his frock-coat handsomely set off a rotund figure, the bulbous nose above, and the crimson face which surrounded it, were less reassuring. He was always a doubtful character, with a tendency to stumble and to smile. But he grew pompous in liquor, and he was pompous now.

'Good day to you, Captain Lawe,' he began. But it was not a hearty salutation; rather a lofty greeting, as from squire to tenant farmer, ringmaster to clown. Something in his birth had given him arrogance, and something in a bottle the need to show it.

'Good day, doctor,' Matthew answered, civilly enough. He would require this man later, good or bad, and it should be in the shape of a willing friend, not a malcontent. 'You enjoyed your noon-time nap, I hope?'

Rushforth looked back at him, and his burgeoning nose rose an inch or two. 'A nap? I do not *nap* in the course of a working day. I have been with my books! One cannot study too much, however great one's experience and skill.'

It was a straight lie, Matthew knew; Dr Rushforth's siesta snores had resounded throughout the ship, and the waking pop of a drawn cork had been as god as a signal gun. But he said nothing. Such fantasies were harmless, if they were overtaken by competent performance later.

Dr Rushforth was staring at the land as if, having discovered it, he was not at all pleased with the results of so great an effort. 'We still run eastwards, I see. Why is that?'

'I wish to examine the coast,' Matthew answered curtly.

'Alone? But what of Captain Downie?'

'By my calculation, the *Blessing* must be a week astern of us. So I can use the time profitably, and then turn back to meet him.'

'But to separate even further?' Rushforth might have been questioning an errant schoolboy. 'Can that be wise?'

Matthew was growing greatly irritated. 'Yes.'

'How so?'

'Because I say so.'

Dr Rushforth, reacting to the stiff tone, said: 'There may be, as we say in my great profession, a second opinion to be had on that . . .' He grew more lordly still. 'I am bound to say, I do not believe my friend Mr Boothroyd would approve it.'

Enough was enough, and the time had come to show it. 'Mr Harkness!' Matthew said loudly. 'A touch on the lee-braces, if you please. The new canvas is still stretching.' Then, when the mate was out of earshot, he turned back to the surgeon.

'Your friend Mr Boothroyd is *not* on board. If he were, he would not supplant the captain. Nor will you sir.'

Rushforth began to splutter. 'There is no question of supplanting. I give you the benefit of my judgement, that is all. You are new to the coast——'

'That is why I take a close look at it,' Matthew interrupted him. 'But I am *not* new to navigation, or determining a course to steer, or estimating a speed, or doing my sums . . . I want no *second opinions* . . . I'll thank you to mind your business, as I shall mind mine.'

Rushforth glared at him. But something of the wind had gone out of the overblown figure, and when he spoke, it was a sulky mumble:

'I sought to be of service . . . As Mr Boothroyd would wish me to.'

'Your service is a surgeon's.' Matthew, once embarked, thought it a moment not to relent, but to drive the lesson home, once and for all. 'I shall have good need of it, before long. Until that time, pray peruse your books, and leave the quarter-deck to me.'

In determined mood, and with all the pleasures of command, Matthew coasted some five hundred miles onwards, to the Bonny River and beyond. It continued with slaves and smoke-signals everywhere, and often a crowd of sails inshore, like ants round a honey-pot, seeking the sweetness of the feast. Then the wind turned against, as his notes of pilotage had foretold, and he put the *Mount Pleasant* about and began to run back.

Fortune followed him, the wind holding in his favour for a

steady westerly thrust. As Captain Downie and his *Blessing* appeared off the port of Whydah, so did Matthew Lawe; and that, with naval navigation supreme, was that.

Anchored offshore within the sight and smell of the port, its fires and smoke, its glimpse of the waiting barracoons, its dark mysterious forests beyond, the two captains met to plan their joined future. Matthew, as the junior, repaired on board the *Mersey Blessing* within an hour of coming into company, and found Downie, to begin with, surly and out of humour.

He sat in his stifling cabin, as airless as a dried-out well, with the customary bottle of Scottish fire-water before him and his bare feet on the desk, and he gave Matthew a look not at all friendly.

'So we meet at last,' he said grumpily, as soon as Matthew was in view. 'I had hoped to see you a damned sight earlier than this.'

'I did not keep you waiting,' Matthew answered. 'Whydah you said, and Whydah it is, within an hour of each other.'

'You did not keep me anything,' Captain Downie snapped back. 'Not company on the voyage, that's certain! Where have you been?'

'I took a turn down the coast. To the Bonny River and back.'

'Bonny? For the love of God! . . . Was that in your orders?'

'Nothing was in my orders except to load slaves on the coast and take them with you to Barbados.' With the best will in the world Matthew was not ready to let this man rule him like a snarling schoolmaster. 'If I have time to spare, I put it to the best use.'

'You would not have *time to spare* if you kept your proper station with me.' Downie took a gulp of the whisky, and seemed to find it as sour as himself. 'Storming on ahead to show off your paces . . . Without me to lead, you might have run into all kinds of mischief.'

'I think not.' Since this nonsense might go on for ever, and waste the day, Matthew sought to mend it. 'Though Dr Rushforth shares your view, and made it plain.'

'*What?*' Downie, startled out of his sulks, took his feet off the desk and sat up straight. 'What has the man been doing now?'

Matthew told the story, as amiably as he could, and was rewarded at last with a change of humour.

'Now 'fore God,' Downie exclaimed, 'that fellow has the gall of Lucifer! I hope you put him down.'

'I did. He has stayed so ever since. Indeed, in his cabin.'

'You may keep him shackled in the slave-hold, for all I care. His trouble of course is liquor,' Downie declared, quaffing again, and deeply, at his glass. 'It gives him courage, and takes his common sense. Now, sit ye down, pour from that bottle—' he indicated the golden-hued Madeira, 'and we'll plan the morrow.'

Comfortably settled again, and restored at least to a state of truce, Captain Downie began a practised monologue.

'There are three ships inshore at Whydah, but they will clear by nightfall. Then we go in to make our choice. The man we see is Da Souza, who is king hereabouts. A Portuguese factor, and knows the trade from beginning to end. He has a compound of barracoons to hold two thousand slaves, and always more waiting to take their place, coming down through the forest or chained to the trees already. He will have sighted the *Blessing* by now, and will keep some good stock for me . . . If we can get our ships alongside, or snug up a creek, so much the better—we embark them directly. If not, they come by canoe through the surf.'

'Is that safe?' Matthew asked.

'No.' Downie's grin was foxy. 'On a bad tide, with the wind blowing south or south-west, it is vilely dangerous. The canoes are often overset. But since the slaves are not paid for until delivered, we do not have to buy shark-meat! So, we make our choice, and wait for them. I always have one of Da Souza's men come on board to collect his price. Thus *he* carries back the gold and the trade-goods, and *he* is murdered for it, not us.'

'Is it so common?'

'As rain in Liverpool! It is dog-eat-dog on the coast, and you should never forget it. We go armed, of course. I trust Da Souza, but he cannot rule every petty chief, every nest of Arab whip-men, every pirate pretending to honest trade. The King of Dahomey himself would take your money, and have a knife stuck in your gizzard, and seize his slaves back again, and sell them to the next fool to set foot on shore! Meanwhile'—Downie rubbed his belly in a grotesque parody of appetite—'who is stewing in the cooking-pot? You! *White meat gives them magic strength!* It is the greatest ju-ju of all.'

'You mean, they are cannibals?'

'They are anything, when they have the chance and the spirit to take it. They would eat your bollocks for breakfast, and the rest for dinner. Man, what a haggis that would make!'

Silence settled in the close cabin as Downie, on a swinish note, came to the end of an evil tale. It was difficult, it was impossible, to believe. Dog eat dog, one might credit. But man eat man? . . . Yet Matthew was left with one feeling above all else, which he had never expected would take hold of him. He was now glad that he had such a barbarous rogue as Captain Downie to guide him.

'Tomorrow, then,' Downie said in dismissal, an hour later. 'Bring Pennyman with you—he is a quick man with the reckoning. And bring your famous doctor too. Let him begin to earn his keep, by helping to judge the flesh.'

The flesh was ready for them, in fearful profusion, as soon as the sailors were led through the stockade and into Da Souza's fine compound of barracoons.

Their march inland from one of the Whydah quays, even though the sun was scarcely up, was a plaguing business. The heat still clung leech-like to the land, as it had done day and night for a thousand years; the host of mosquitoes welcomed this new blood, and swarmed towards them like orphan children at a picnic. From the rain forest surrounding the compound, the steamy smell of Africa, and soon of Africans themselves, came wafting down in great gulps of fetid air.

This forest was not still; it was alive with moaning, with lamentation, with pain. The sound of slaves taking a last farewell of their homeland, or still raging and wrestling against a pitiless separation, was the sole dawn-chorus of Africa. In this tormented corner of the world, loud with the rape of innocence, no birds sang.

By the time the motley band from the *Mount Pleasant* and the *Blessing*, three men from each ship, entered the compound, they were drenched with sweat, and itching from a dozen bites and stings. The surgeons had already done their best to armour them all against these, with sulphur ointment for limbs and a draught of quinine bark for the bloodstream. For all the protection these afforded, they might have been the mosquitoes' favourite feast.

But after the march there came, for a short time, the respite of hospitality.

The visitors were met at the door of a thatch-roofed *rondavel*, balconied, white-painted, of some spacious elegance, by the king of this empire, Da Souza the great slave-broker. He was a middle-aged mulatto, splendidly strong, attired in a gossamer silk frock-coat which could not have been matched on this side of the great Paris *faubourgs*. His eyes were watchful, his manner confident, and his greeting gracious as he first welcomed Captain Downie as an old friend, and then Captain Lawe as a newcomer to the coast.

It was cool within the *rondavel*, aided by swinging punkah fans of green palmyra leaves which were kept in motion by the best machinery of all—a trio of tireless house-slaves. It was cooler still when, after polite inquiries as to their voyage, Da Souza dispensed a delicious brew of lime-sherbet in a stone joram drawn up from the well in the centre courtyard. It was cool and calm as the gardens of paradise, until the moment when the slave-king said:

'Well, you are here on business, Captain Downie. I'll not detain you from making your choice, and then we will talk again. What are your requirements?'

'Perhaps one thousand and three hundred, between the two ships,' Downie answered.

Da Souza raised his eyebrows, though in courtly fashion. 'You have space?'

'We will make space.'

'I would not dream of doubting it . . . I set aside some nine hundred, prime stock, no idle bones or sick, as soon as your two ships were sighted in company. But four hundred more will not strain my larder . . . Let me now deliver you to the good offices of my overseer. A new man, Barbosa, but not new to his task. You know that you may ask for any possible help within his power. Until later!'

Then it was time to march from the cool into the heat again, and so to the barracoons, where all elegance and ease shrivelled to nothing on the gross, deformed branches of this tree of despair.

Barbosa the overseer, a tall gaunt man with ringlets of yellow hair and the look of fever about him, carried a whip: a weapon markedly more handsome than himself, being a long-tailed switch of plaited horsehair, with a golden top and a feathered end.

It was his badge of office, but it served for more than show; he enjoyed the use of it, and as the party entered the first barracoon, led by this man of power, there was scarcely a slave to be passed who did not receive a sharp cut, anywhere on his naked body, from buttocks to neck, from the small of his back to the tendon of his heel.

If the victim, being brave or insensible, failed to yelp with pain, he promptly received another as a reminder of his station.

The first barracoon, and the five others which made up the slave compound, were thatched with palm leaves, like Da Souza's lordly quarters, but there all resemblance between the estate of master and man ended. They were simple sheds, like the long-houses of a thousand African villages: fashioned of tree trunks driven into the soft earth during the rainy season, boarded and lashed together with tendrils of liana, stark, useful, and forbidding.

Lengths of chain, some five feet from the floor, were slung from one end to the other: chain marked at intervals of a yard or less by a neck-link, into which each inhabitant was locked. He was secured in such a fashion that he must stand upright, his feet on the planked floor, his drooping body fully exposed. There were women in the barracoons, but not bound, only caged. There were children also, silent, staring like cats terrified from their wits, herded in the same fashion.

The slaves were separated, as Barbosa, his whip always busy

with discipline, cheerfully explained to Matthew, into marketable lines of merchandise: full male, full female, men-boys, women-girls, boys, girls, women nursing, women carrying—and the unfortunate old.

Over all, under all, from them all, there rose a stench of fear, sweat, and excrement so forcible and vile that the very dogs of the compound kept their distance. But surveying the throng with gleaming eyes, Captain Downie had only one comment to make. Rubbing his hands together, he exclaimed:

'This will teach the Evangelicals!'

Downie had with him his own doctor and trading mate from the *Mersey Blessing*, the one a student of Edinburgh who had put to sea for his health—not, perhaps the best recommendation for a medical man in charge of so many other suffering souls—the other a counting-house clerk who was still learning the ways of infamy.

The two parties divided, and when an hour later they met again in the fourth of Da Souza's barracoons, they could agree on one matter—that the nine hundred slaves set aside for their purchase were as the slave-master had described them: prime stock, well worth the buying if the price was fair.

But now another chapter opened with the arrival of a further four hundred slaves, those whom Da Souza had declared 'would not strain his larder'. Sharp-nosed Pennyman had already warned Matthew that these might be suspect: if they had been kept chained to the trees instead of being displayed in the barracoons, it might be for the good reason that they were not first-class merchandise. So it proved.

The late arrivals were chased and herded into an empty barracoon by Arab whip-men who (as Barbosa phrased it) had already seen enough of their damned black hides on a month-long journey from the dark interior, and only wished to be rid of them. They were the very dross and ordure of the trade: old men, wounded young warriors, malefactors who had been steadily whipped into the posture of curs, moon-faced children whose brains would never match their coming manhood, even if they were spared to reach it.

Barbosa the overseer, whose switch was now most active, was

hard pressed to persuade the buyers that their money would not be wasted.

'Now,' he said at one moment, 'here we have a fine lot of five. Thin, because of poor rain. But they will gain.' He indicated a wretched crew of five young men who might have been brothers, and were certainly brothers in misfortune: exhausted skeletons, broken in spirit, beyond caring, almost beyond mercy. It would have needed God's own magic wand to make them 'gain' anything but an early grave . . . 'You will take them?'

After a long silence Captain Downie pointed to one unfortunate, and growled:

'Make him hop!'

'What?'

'You know well what I mean. Let me see his paces.'

Unwillingly, Barbosa applied his whip, with a single upward flick between the shrivelled thighs. The young man—now, in spirit, infinitely old—jumped in the air, came to earth on what seemed a tortured foot, and rolled down with a scream into the dust and foulness of his prison floor.

'I knew it,' Downie said, mightily pleased. 'He has the jiggers!'

'What is that?' Matthew asked.

'The Guinea worm. It burrows into their feet. It rots the sole, and will rot the sale.' He laughed, well pleased with his jest. 'No,' he said to Barbosa, 'I'll have none of him, nor of the rest.'

'What of the children, then? Good teeth, good growers. They could be cutting cane by next summer.'

It was the turn of Dr Rushforth. Rushforth was by now plunged deep in suffering, from the heat, the mosquitoes, the simple torture of walking: he had the look of a man for whom a glass of cool sherbert was insufficient answer to his woes.

'Too young, too small,' he declared, as a Roman emperor might have decreed death to a laggard gladiator. 'Send them back to the trees. Let them put on a year's growth. They are not worth *our* feeding.'

The choice continued, in a climate of ill-humour. Mathew,

who could do little but watch and learn, was left with one strange remembrance. Though all the grown men were stark naked, when they were being examined, and prodded, and their muscles felt, they always covered their private parts; and when intrusive hands pulled this mask aside in search of the venereal, these outcast savages moaned and hung their heads in shame. All other miseries seemed to pale beside a gross indecency.

They had been schooled, from childhood, in modest behaviour, and there were young girls in the compound.

Returning to Da Souza's fine house, the buyers found a fine welcome. Dr Rushforth received at last the blessed sacrament of a bottle of cool Portuguese wine, Downie that glass of whisky which must by now have become a tradition, and Matthew, after hospitable inquiry, his Madeira. Thus lubricated, and at ease, their bargaining began. It was close work, as all parties had expected.

Da Souza, enthroned in his wicker-work chair with a canopy above his head, made the first move. He was affable as usual, but his face had grown sharper, as if the smell of money was drawing all his features to a point, like a questing hound fixed firmly on the scent.

'You will take your thirteen hundred?' he asked Downie.

'No,' Downie answered, with nothing in his voice save a cool rejection. 'The nine hundred you set aside for us—and I thank you kindly for the service—are good stock, saleable, though most of the older men are past their best, even for kitchen boys. But the rest—well, let us say that the sharks would welcome some of them more than I . . . We will take eleven hundred. The two hundred new have all been marked with cochineal dye. Barbosa is chaining them with the others.'

'Eleven hundred.' If Da Souza was disappointed, he would not show it. 'Well, I will not lose such good friends for two hundred slaves. Let it be eleven, then. Shall we set a price for the whole cargo, without distinction?'

'Aye. If the price is fair. 'Twill save time.'

'And money too, if I know you, captain.' Da Souza smiled his last smile, and turned businesslike again. 'First, there is my own price in gold. The market has risen in Barbados, as you

60

know. There, they are crying out for slaves . . . Shall we say, two pounds a head?'

Pennyman the purser, whose finest hour of enjoyment this was, allowed himself to draw his breath with a sharp hiss of alarm. Even Captain Downie's trading mate, new to the game, copied him, and pretended to mis-swallow his wine. Downie translated this ploy into words.

'We may *say* it. We will not pay it. Two pounds! Do prices double themselves in a year?'

'Yes,' Da Souza answered, with flat determination. 'And again next year, for all I know. The increase is there, for all to see.'

'What increase is this? I know no increase. Have the slaves grown bigger? They did not look so, this morning. Some of them had shrunk down to little bags of bones!'

'The increase at auction. You know it, and I know it. We do not live in ignorance, because we live at Whydah! In Barbados, on the auction block, you may get forty pounds for a strong working buck.'

'Aye, very likely,' Downie agreed. 'And five pounds for a skin-and-bone child. It is the average that rules, not the price of one buck at his best, either working or breeding. And how many of our eleven hundred shall we land alive? You know the wastage.'

'I do not know the wastage,' Da Souza answered, pretending to be near disdain. 'That is a matter for you and your surgeons. *I* sell, *you* carry. If *I* sell them prime, and *you* land them rotten, am I to pay for that?'

'No.' Downie was still cool. 'We each take our chance. That is why the price should be the same as a twelve-month ago. One pound a head, the good and the bad together. You have the money, we have the trouble. It has always been so.'

'I know it well enough! So you are rich, and I am poor.'

'My dear sir,' Captain Downie said, suddenly affable, looking round their splendid quarters and then directly at Da Souza's gleaming suit of silk, 'say the word, and I will change with you.'

So they continued to bargain and to chafe each other, the

Portuguese against the Scotsman, the seller against the buyer. After a half-hour the price was settled, perhaps at what had been privately determined in the first place: one pound and ten shillings a head, in gold coin, the rough with the smooth and, for the slaves, the living with the dead. Pennyman, the great accountant, read out the terms of their hard-fought agreement from his note-book.

'Price to Mr Da Souza, for eleven hundred slaves as selected, male and female both, one pound and ten shillings *per capitum*. Total in gold of the English realm, one thousand, six hundred and fifty pounds.'

'Agreed?' Downie asked.

'Agreed.'

The word at last was enough. Even in this hyena trade, there must be a pause for honour, as the spoil was divided. But there were still many such rounds, as between prize-fighters who lose one and win another; the sun had already drawn westwards, bringing ease to a burning sky, before the bargaining was done and the word 'Agreed' was spoken for the last time. First to be rewarded were the slave-runners, Da Souza's own men, who had driven their prey down through the forests and, paid once by their master, expected gifts of good English quality by way of 'dash' to seal the bargain.

For them there must be swords, and mirrors, and gunpowder, and bolts of silk, and buckled shoes, and cases of bottled ale. How many? Of what value? Both ships had a good stock of such vanities, but wished to waste as little as might be on these Portuguese land-pirates: there was a ready market waiting for the surplus in Barbados, to be bartered for rum and sugar, the last profit of the voyage.

After the slave-runners there were still the petty chiefs of the interior, the betrayers of their tribe, whose greed must be assuaged by mountainous tons of trash goods from other dark interiors—Manchester, London, Birmingham, Liverpool. Here the currency was either gaudy or useful: beads, necklaces, bangles which had the look of gold and the worth of brass, nails, fish-hooks, knives: clockwork toys, chiming bells, straw bonnets, tin trays bearing the likeness of King George of England.

Once again—how many, and of what value? Such men had been paid once for their treachery, like the forest-runners; now they must be flattered and cajoled, to promise a future of equal profit, equal readiness to betray.

If they were pleased with their beads, they would answer with flesh, until the flesh itself was exhausted.

Pennyman, busy as a wasp at a picnic, conferred with his captain, sought approval from Downie, matched wits with Da Souza, did his formidable sums, and delivered the final accounting:

'Case-goods of quality, eighteen. Trade goods, ninety chests of one hundredweight each. Value certified, two hundred and forty pounds, English.'

'Agreed?'

'Agreed.'

It was done.

At dawn next morning Harkness the mate turned up all hands on board the *Mount Pleasant,* and set them to the task of which he knew all that was needed to be known: preparing a slave-ship for her proper cargo.

The brig, under Mr Benjamin Boothroyd's all-seeing eye, had been carefully designed for this transformation: the innocent cargo-carrier could become the serviceable floating dungeon in the space of half a day, and then altered back again within a week—the generous extra lay-days being necessary for cleansing the ship.

First the case-goods and the trade goods, more than one hundred chests and packages in all, were brought up, hoisted ashore, and set down upon the Whydah quay with an armed man to guard them. He had only one order: 'In case of thieving, shout the alarm, then shoot—at black man, white man, mulatto, dog, ape, child, or elephant!'

Such a position of honour was much coveted: it carried a dram of rum every hour, no heave-and-haul like common sailors must endure, and the admiration of children, who recognized a king of his own realm when they saw one.

Then came the preparation of the slave-decks. This was carpenters' work and, in the fierce heat of morning on the Equator, the most wearisome of all their voyage. The empty cargo holds were now sliced in half like rounds of beef from one end to the other, with struts of timber and planking to divide the layers: so that what had been two stowage spaces each six feet high were now changed to four separate decks, whose headroom was a precise three feet.

As the meat of these meagre sandwiches the slaves would lie there chained for the journey; and there they would endure, eat, drink, sleep, and excrete, like manikins in the filth of their own aspic.

Lastly, the endless lines of chain and shackle were rigged for the males, and a cage built for the unbound women and children; and thus was this great hotel of the sea made ready for its guests.

Some three hours after noon, when all the sweating work was finished in both ships, Captain Downie came aboard for final conference. He was in better humour than Matthew had expected, after such a toil; and when he had looked round the *Mount Pleasant*, and approved her, they stood together in something like amity under the awning of the after-deck.

'The cargo is due to be delivered now, as you know,' Downie said. 'We should hurry them below, and stow them as smartly as we can. I want to catch the evening tide, before some Navy ship comes sneaking in to trap us.'

'Would they do so?' Matthew asked. He did not like the phrase 'sneaking in' when attached to the Navy, but he let it go. 'What is the position in law?'

'The position *in law*,' Downie answered, with sarcastic emphasis, 'is the same as the position in fact. By the gospel according to Saint Wilberforce, whom God preserve until he fries in hell, it is this: that if we are loaded, we are guilty. We are condemned as a slave-ship, and can be cursed off the face of the earth. It would be best if we are clear of the coast, and gain plenty of sea-room, before such an evil fate befalls.'

He took a turn across the deck, glanced aloft as did every captain, whether in his own ship or a stranger's, and then returned. Matthew noted his nervous state, and knew that there was something in this moment of balance which was, for

Downie, the great challenge of the voyage. He could not like such a man, but one might profit from his experience. He asked:

'When we are at sea, what rules do you keep?'

'Rules?'

'For the regulation of the slaves.'

'There is only one rule,' Downie answered forcefully. *'Keep them down!* They will kill you else! So. do not let your blackbirds out on deck at all, unless there is fever, and they must be cleaned. In that case, give them freedom only in small groups, not more than twenty at a time. You will be armed, as I told you in Liverpool. But see that your gunner has all his smallfire weapons ready. And be ready to use them. Better a score of dead slaves than one dead captain!'

'But if they are chained——'

'They have been known to slip their chains. Sweat and short rations can loosen a manacle. They may even feign death, then jump up and hack the burial party to pieces. Oh, they are clever devils, and do not forget it! You need not cosset them either. It will not make you friends, only strong enemies. One boiler of rice a day will feed ten men. When we are two weeks from Barbados, that will be time enough to flesh them out a little. Your doctor knows the best means of that.'

'What are these best means?' Matthew asked.

'Beef for muscle, and cod's liver oil to put a shine on their hides. It is not only ladies' maids who dabble in cosmetics! You have both beef and the oil on board. Make them take it! And one more matter—keep their nails well cut. They can kill each other fighting.'

Matthew was startled. 'They fight *each other*?'

'Aye. It is tribe against tribe, even in captivity. What can you expect of such savages?'

What indeed? . . . Matthew felt that he was now wiser, and he tried valiantly to suppress any other feeling. If he was in the slave-trade, then he must excel in it, and forget all the rest. He said:

'So it is keep them chained, keep them fed enough, and guard them closely?'

'Aye. That is the size of it . . . If they seem threatening and might somehow break out, you may strew some broken glass on the deck outside the slave-holds. 'Twill keep them at bay till you have time to shoot, or put on the whips.' Suddenly Downie raised his bony beak and sniffed the heavy air. 'By God!' he said coarsely, 'you can nose these damned blackamoors a mile away! They must be coming.'

They were coming indeed.

Whether Downie had in fact smelled the honest, terrified sweat of their cargo, or had seen a stirring among the trees and wished to pretend some special skill, Matthew could not decide. All men had their dreams of wizardry; perhaps Downie's was this magic gift of 'nosing a blackamoor a mile away'. But his prophecy came true, within minutes of his making it.

There was first a general, mournful sound, as if of a lowing herd of cattle driven near the limits of their strength. Coming from human throats, as it must, it was desolate indeed: an anthem of despair closer to the gates of hell than the courts of heaven. Then there was movement at the edge of the clearing which contained the Whydah quays; and then, with a louder howling and groaning, the dark lines of the slaves emerged from the forest and began to shamble towards the ships.

Exhausted by the steamy heat of the afternoon and by weeks of forced marches through jungles where the most cruel animals were still called human, beset by much whipping and cursing from the Arab slave-runners whose last chance of domination this was, they still had strength enough to cry out their misery. Some of this chorus was mere whimpering, of men beaten nearly insensible for trying to escape; others howled aloud for their lost families, seized from them and taken by another ship to another land.

When they sighted the waiting ships, it was as if they realized at last their true fate: they must now be ripped from their homeland, even as some had already been torn from their wives and children. The groaning and howling increased, and now the women joined in with wild shrieks and the children, catching fear from parents suddenly brought so low, cowered like the wild young of animals and hid their eyes as if they could thus hide their bodies.

Then the endless lines, the multitude of the lost, were marshalled on the quays; and the first of the *Mount Pleasant*'s cargo began to crawl on board, and mighty Boatswain Pugh came into his own.

Pugh, a Liverpool man with Welsh forebears, had outgrown his dark ancestors to sprout red hair, shoulders like those of a southern shire-horse, and the immutable belief that big men were better than small men, and very big men the best of all. He had retained only a spiteful temper to link him with a race which said Yes to themselves, No to foreigners, and Maybe to possible friends.

Under the benevolent captain of the *Mount Pleasant*, his rule had so far lacked either tyranny or brutality; now his ordained subjects were coming aboard, and only a soft fool would stand in his way. With ready whip and forceful sea-boot he began to dispose of them, while Harkness, and Dunhill the second mate, supplied the brains, and he the muscle, and Matthew Lawe the uncertain conscience of mankind.

By agreement, Captain Downie was to take five hundred of the slaves, and the *Mount Pleasant* six hundred; the stowage of this miserable flesh, though less than what had been planned in the first place, was still a tight fitment. The ship resounded to the clanking of chains, roars of command, thud of hammer on shackle, cries of pain, wailing prayers for mercy—all the common evidence of slavery in action.

At one moment, a certain desperate man made his last bid for freedom. Half-way up the gang-plank, crawling in humble liberty between one set of chains and another, he turned and fled ashore. He was an agile fellow, and eluded both the guards on board and those who had just released him. On the quay, he found himself face to face with the man protecting the landed goods, which were the price of his own slavery.

This sentry shouted the alarm and raised his gun, but he did not fire; a narrow view of his duty determined that the fugitive was not a thief but a runaway, a thing of living value only, and thus the quarry of someone else. The slave, in howling despair, turned and dived into the river.

Matthew watched in utter sickness of spirit as the bobbing black head began to make progress from one side of the Whydah Creek to the other. But for such an outlaw there could be no escape: he was only swimming from his second prison to his last. As he neared the opposite bank, guards of his own colour ran down to the foreshore, whooping and shouting, and made ready to intercept him.

Forlornly, the tiny figure ceased to swim, and in hopeless surrender merely floated with the ebb tide towards the open sea.

Boatswain Pugh pronounced his verdict:

'The sharks will take him soon enough. Good riddance!'

Pennyman the trading mate closed his own account in his own fashion:

'He was not on board,' he declared roundly. 'He was not yet ours. He was nearer to the quay than the ship. *Not delivered*, he pronounced, and with a firm hand scored out one soul from the total of the elect.

Down below in the slave-decks, the work went on. It was never pretty, and often barbarous: treat a Jew of Liverpool in such a fashion, Matthew thought, as a rebellious slave, axle-greased with his own excrement, was rolled into his appointed place, and even pious Christians would come running to his aid.

Under three feet of headroom, with the masters forced to creep and crawl over their prey, it was push, shove, whip, manacle, and then forget, as these sardines of men were forced into compact rows, to make their last exchange from humanity to merchandise.

They must go into their allotted space, like any other fish salted down into the hold . . . Presently it became clear that the cargo of the *Mount Pleasant*, six hundred of his gasping land-harvest, was doomed to overflow, and could not be parcelled up. For all the careful planning in far-off Liverpool, square feet of flesh had swollen to a baffling schoolboy sum, multiplied by blood, divided by bone, and not to be solved in terms of square inches of space.

But Boatswain Pugh, the master-packager, saw no problem here.

'If they will not cram in one way, then they must do it in another,' he growled to Harkness, who was prepared to say No, and recommend a lesser cargo. 'Six hundred into four deck *will* go, if we dress it aright . . .' He flexed his mighty

Side view and Plan of a Slave Ship

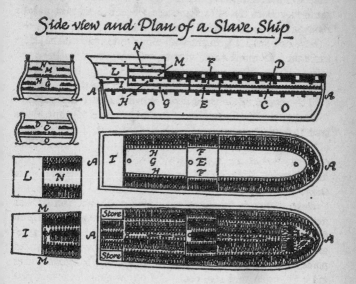

muscles and glared across the mutinous tumbling sea of legs and arms and rolling eyes. 'Turn 'em sideways! We have done it before. 'Twill save six inches for each man. 'Tis their own fault for sprouting those big black bottoms!'

Their own fault decreed their own punishment. Painfully, with more whipping and kicking than had yet been used, the new stowage was done. Each captive was turned over on the shoulder opposite to his heart (Dr Rushforth's sage advice, 'to liberate the bloodstream'); since their heads were placed outboard against the ship's hull, those on the starboard side faced their future, and those on the other gazed backwards at their vanishing homes. None could see either.

Now they lay close in each other's laps, like spoons in a wedding chest of silver. Their nails were cut, their mouths doused with vinegar by the same prudent doctor; and, in such prime condition, they were prepared for export.

Towards dusk, when all was done save to conquer the stench and to silence a clamorous moaning which could never be stilled on the feeling side of the grave, Da Souza's black porters came to collect the chests and cases which were now due to him, and his confidential clerk to count and bear away the golden guineas. He brought with him a letter from his master, flowered with many compliments, and ending:

'I wish you, in your own words: "Many happy returns!" '

Well-slaved indeed, and in good time to use the last of the ebb, the two brigs dropped down Whydah Creek and out into the open sea. Matthew, always as glad to sail as to return, was glad no more. Between one sunset and the next, the sweet smell of a new ship had turned foul for ever.

4

One week later, off the Grain Coast south of Sierra Leone, already a haven for freed or runaway slaves, the *Mount Pleasant* and the *Mersey Blessing* were still in company, three degrees north of the Equator and eight hundred miles west of Whydah, and ghosting along under light airs towards the main Atlantic. Their voyage so far had been without incident, save for a killing in the slave-decks and one woman dead in childbirth.

Only their convoy of sharks, cruising close alongside and stretching far astern as an evil wake, gave the lie to such peaceful progress. A cargo of misery, now smelling to high heaven and beseeching it in vain, continually signalled to these escorts and to any ship within five miles of their stench: 'Here are slaves. Come buy! *Come eat!*'

Then, on a certain day in July, the two ships, becalmed under the sulphurous heat of noon, found that they had distant company.

It was topsails first, far to the north-west of them, glimpsed above the hazy mirror of the sea, taking shape like little clouds approaching. They were clean topsails, full topsails, moving gently towards them, seeming to carry their own charmed wind within their embrace. From the *Mount Pleasant*, drifting sluggishly to the south of the *Blessing*, Matthew hailed his fellow captain across a glassy stillness which scarcely needed a speaking-trumpet to overcome its interval, and received a grumpy answer:

'I have eyes, thank'ee kindly!' Downie shouted back. 'I can see topsails.'

Matthew, nettled as any man might be who had essayed a friendly warning under equatorial heat, and got a Scottish bagpipe up his nose by way of return, thought it suitable time for a little sport.

'Can you see *Navy* topsails?'
—'Eh? Say again.'
'Can you see Navy topsails?'

There was a change of tone in Captain Downie's disagreeable voice as he asked: 'What mean you?'

'They are *clean*, Captain Downie. They are *white*. Take pride in them!'

Then Matthew retreated into the shade of his own mainsail, and left a morsel of poison to fester as it might. He was doing no more than making a guess, but the perverse pleasure was not the less for that.

Within two hours he was proved right. The clean topsails became a clean spread of canvas, on three serviceable masts, and underneath them a single checkered pattern of gun-ports. One tier . . . The stranger was a small ship-of-war. The stranger was a British ship-of-war. The stranger was a British frigate, slowly bearing down upon them with a wind and a will of her own.

It gave him great pleasure that Captain Downie was the first to break an unfriendly silence. After he had hailed, with something akin to naval formality: '*Mount Pleasant*, ahoy!' Downie went on to ask:

'Can you make her out?'
'Aye.'
'What then?'

Matthew, whose telescope had been busy, answered: 'A Navy frigate. Twenty-eight or thirty-two guns.' It needed no voice from the grave, no ghostly hail from the battlements, for him to add: 'They have sighted us.'

'Well, God damn them! What next?'

But Matthew was not ready to help this other, closer enemy. He had his own preparations to put in hand. He made a second guess, which might be either fatal or profitable. If they were becalmed, while the frigate carried her own northerly wind, then there would be nothing surprising if such a chance breeze died, and another from the south, which usually prevailed, took its place.

It was worth a wager . . . If this sea-constable caught them full of slaves, it was worth *any* wager . . . Calling Harkness the mate, he had him trim the yards back, and set the wheel, so that this southerly gift, if it came, would catch the *Mount Pleasant* first and pull her clear. Then, ready to disengage, he waited.

He waited—they waited—five more hours for the frigate to close. After watching her anxiously for much of that time, Matthew began to play a foolish game which he prayed might have some magic in it. He would deliberately turn his back on the approaching ship, and gaze southwards for ten or fifteen minutes at a space: then turn again, in the hope that her northerly wind had dropped and died, that she had not grown any bigger nor come any nearer.

But the magic and the foolishness for ever failed him. She was always bigger, always nearer, and her wind never died and the south breeze never came, and before long the torpid brigs must be caught.

At six o'clock in the evening the warship was less than eight miles away, but moving more slowly. Matthew, gazing through his telescope, noted everything about her that might be of service. Though her canvas was spotless, fresh from the sail-lofts of Chatham or Plymouth, she was an old ship, with a scarred hull and bulging gun-ports of a shape long out of naval fashion. She lay a little high in the water, which might mean she was short of provisions, or guns, or even men. She steered true as an arrow. Her figurehead—

Matthew's heart seemed to stop suddenly. He had recognized this ship, with a sense of shame which flooded him throughout his body, from throat to stomach-pit. The figure-head, a curious carving of a hawk with a wreath of laurel in its beak, known to lewd sailors as the Hole and Pecker, was one which he had last seen nearly thirty years before, on the West Indies Station where he had been one of her proud company.

She was the old *Solebay* frigate, 32 guns, a ship of honour and repute, and they had then been fighting the same battle and serving the same sovereign.

Sick at heart, he watched her moving closer. How could he treat her as an enemy? . . . Then, as if time had been called by some merciful ringmaster of the sea, her carrying wind died away to nothing, and the three ships sat motionless in the same fatal line, like the hands of a clock on the stroke of six: the *Solebay* some five miles away, the *Mersey Blessing* between them, and his own *Mount Pleasant* half a mile to the south, on the hour-hand itself.

Presently there was a sudden flash of flame on the frigate's foredeck, and then the sound of a shot: a blank cartridge from her bow-chaser, a warning to behave and conform. Then Matthew ceased to watch the *Solebay*, and gazed instead, with increasing horror, at what was happening on board the *Blessing*.

For a long time he had been aware of great activity in Downie's ship: the sound of shouting, a continual clanking of chains, a sort of creeping tumult of people at work on the side nearest to him, and masked from the frigate. Then one of her loading-ports, close to the water line and never to be broached at sea, did indeed swing wide open, and with a rattle and a roar a great length of cable ran out, and down, and plunged into the water.

To it were manacled, at close intervals, dozens, then scores of writhing black bodies. Captain Downie, convinced of danger, was making jetsam of his cargo.

Matthew, aghast, noted with final revulsion that all his company of sharks had now deserted him, and were swarming round the other ship. They would have to be quick . . .

Harkness was by his side on the quarter-deck, and Dr Rushforth. They were watching with equal attention, but their feeling did not match his own. When he muttered: 'What a foul thing to do!' Harkness only answered:

'Aye, sir, foul enough. But it will be fouler for him if he is caught with his blackbirds on board.'

Rushforth was almost magisterial as he gave his judgement. 'Better to lose five hundred slaves than forfeit the whole ship. It is common prudence.'

Harkness looked at his captain. 'What do we do, sir?'

'Not that, by God!'

A second chain laden with doomed, screaming captives began to run out from the *Mersey Blessing*, like a flux of bloody iron ordure escaping. Common prudence was at work again.

Matthew, ready to look anywhere save on this frightful picture, once more raised his telescope to the *Solebay*. Though she remained utterly becalmed, she was not idle; she was now lowering a boat—two boats, in which the glint of fire-arms could be seen. A boarding party was on the way. It would be a long pull, but sailors would do it.

Then fate rescued him from an appalling, hopeless dilemma, to which, for the very salvation of his soul, he had been ready to surrender. The creaking of spars above his head, and the brisk slatting of canvas, made him glance upwards. Their sails at last were stirring. He swung round to look southwards, and there was the expected ruffle of wind on the surface of the water. They had their southerly breeze, in the very last pinch of time!

The *Solebay*, old and slow, with two boats lowered, would never catch them if they moved quickly. Dusk would come within the hour, and then the black domino of night . . . He shouted to the helmsman: 'Down helm! Bring her round south by west!' and then to Harkness: 'Call all hands. Brace those yards back till they crack! We have two hours' start on them, and by God we'll use it!'

There was only one morsel of relief in this ghastly scene of shame. He could run from it, as fast as God allowed.

God allowed not only escape, but a most merciful relief, signalled in matchless colours, warm breezes, and the finest of all benedictions, the freedom of the sea. The Middle Passage of the Atlantic, which could turn brutal at any season of the year, now proved to be kind—kind, at least, to all those licensed to move above decks.

The *Mount Pleasant* had slipped into the dusk, and then donned the cloak of night, while proud old *Solebay* was still powerless to stop her. The frigate had fired a few vain shots which could only fall short, and then decided, after these growls of disapproval, to continue boarding the *Mersey Blessing* and to let her consort go. Next morning, Matthew awoke to a solitary, pale yellow dawn without a sail in sight, and then, turning his ship westwards for Barbados, to calm seas and a prosperous voyage.

Thus saints and sinners had been sorted out, and Matthew Lawe was, to his astonishment, numbered among the elect.

The voyage could not be prosperous for all mankind. Below in the slave-decks, there were fights, there was foul sickness, there were deaths which could only be set down to mere despair. The women, though sisters in misery, took to fighting among themselves, and it would have needed a company of sailors very brave indeed to stand between one shrieking band of black Amazons and another. More prudent, it was thought, to wait, and bury the defeated dead . . .

Dr Rushforth expressed his disapproval in sanctimonious terms:

'To think that we have rescued these savages—*ransomed* them, indeed—from the clutches of their own harsh rulers, only to find them so base, so ungrateful!'

But there came another day, towards the seventh and last week of their voyage, when Dr Rushforth proved that he had qualities far above such pompous piety, and at last justified his keep.

He had entered Matthew's cabin, a little portly man, waddling like King Duck of the Farmyard, and confessed to some anxiety.

'I visited the slave-decks with the feeding-party this morning,' he began—and indeed, a certain aroma round his person confirmed such an excursion. 'All is not well, Captain Lawe, and it is my duty to tell you so.'

'What is not well, doctor?' Matthew, as much at ease as any ship's captain might be when granted fair weather, a stout vessel, and an easy soldier's wind, could not truly cavil at such small mosquitoes as Dr Rushforth and his prosy pronouncements. But there was always perfection to be yearned for . . . 'Is it the women again?'

'No. They are subdued, thanks be to God. It is sickness. Very severe. We shall surely lose slaves if it is not mended.'

'What sickness?'

'A loose running flux, with mucous blood evacuated. The worst I have ever seen. It is a contagion, like the plague, and can only gain if they remain in their filth below. Many will not eat. They *cannot* eat. They vomit up, or burst their bowels below.' Dr Rushforth consulted a piece of paper. 'From all causes—disease, accident, violence—we have lost twenty-six of the cargo already. In a week of this pollution, we might lose ten times that number.'

Matthew was greatly disturbed. The prime rule of the Middle Passage— 'Avoid wastage of cargo'—was in danger of being breached, and might bring their voyage to ruin.

'What's to do then?'

'I can treat them,' Rushforth declared importantly. No malodorous talk of vomit or bursting bowels could affect his measured style. 'But I cannot treat then *in situ*—or, to be more precise, while they are anchored in their excrement. They should be brought on deck, perhaps a hundred at a time, so that I may go to work in proper circumstances.'

'What work will this be?'

'Nature's cure, and mine! Give them a change of air, and a chance of sunshine. Give them a little heart. Wash them all down at the pumps. Consign to the deep those who are too far gone, whether alive or dead: they will only poison the rest. Then a distribution of cordial, with lime juice or spirits of wine. We have the means to concoct it. Then some meat with their rice. Eggs if we have them. Eggs are good binding stuff . . . While they are out on deck, douse their quarters with vinegar.

Burn plenty of brimstone, to smoke out the evil. It will make the rest of them cough abominably, but better coughing than coffin! Eh? Eh?' Rushforth looked towards his captain for appreciation of this subtle jest, found none, and continued unabashed: 'We cannot cleanse those decks, this side of Barbados, but we can render them less foul. I believe——'

Matthew interrupted a prescription which might continue for ever. 'I see nothing in the way of that,' he said.

Rushforth pursed his little mouth. 'Others do, captain. I can assure you of that. Harkness says No, the slaves should not be on deck. Boatswain Pugh says No, more forcibly, as is his habit. The gunner says No—he has not the small-arms to guard a hundred men at one time. But in my opinion'—would he never lose his sermonizing tone?—'it is *your* plain duty to say Yes.'

Matthew, though irritated as always by this pedant who would now teach him his 'plain duty', came to his resolve. He had not visited the cargo decks since the stowage of the slaves at Whydah: Harkness and Pugh had seen to their discipline, and Rushforth to their health. It was not quite enough, for a conscientious captain with, perhaps, over-delicate feelings. He told Rushforth:

'It is not for them to decide. With your help, I will see matters for myself.'

Dr Rushforth was startled. 'You mean, go down to the slave-decks?'

'Aye.'

'They are foul beyond description, Captain Lawe.'

'Then they must be cured . . . Kindly tell Harkness and Pugh that I will come down in ten minutes exactly.'

'If I may offer advice—'

'What now, doctor?' Matthew interrupted sharply.

'It is in the matter of dress. My own is always *old* boots, *old* breeches, and a stout sheath of sacking which may be thrown overboard as soon as may be.'

Matthew was tempted to say: God bless my soul, I was thinking of gold-buckled pumps and satin trews! but he forbore. The man was doing his best . . . He dismissed him, and made himself ready in serviceable style.

What he could *not* be made ready for were the slave-decks themselves. 'Foul beyond description' fell ten leagues short of reality. With hopping little Harkness and bulky Boatswain Pugh in attendance—the latter sullen as he dared to be, as if at some usurpation of authority—he followed Dr Rushforth down to the first of the holds, passing the caged women who, knowing they would not be whipped, shrieked abuse and clawed like black harpies at the visitors.

But after that, in the eerie darkness and choking foetor of the male decks, there was no such spirit left.

Matthew found himself shuffling on his knees through a stinking tide of loose excrement, writhing feet, and spilled rice which not a starving rat could have swallowed. A few men still wrestled with their chains, and the rattle and ring of iron mingled with a constant moaning and howling, to make the very music of the damned.

A few others lay in stupor or in death. The green-backed flesh-flies which had followed them in swarms from Whydah were now feasting. A huge Negro wrenched his body upwards, lunging at Matthew as he passed, and then dropped back as if this had been his last movement on earth. Brave man, Matthew thought, as Pugh lashed out at the still body, and was answered only by silence.

The moaning and the clanking persisted, the stench of this prostrate congregation grew until it stupefied the nostrils. Matthew, having crawled along one deck, found that he could go no further. He was retching, and he already knew enough . . . Their party retreated, to find the clean air of the upper deck almost as stunning as the pit-vapours below.

Matthew stood by the rail, stripped off his sacking gaiters, and cast them swiftly overboard. Then he sluiced his hands, and the soles of his boots, in a bucket of sea-water. Even so, it seemed that the foulness would never leave him. He said to Harkness:

'They must be brought on deck, in batches of one hundred, and cleansed.'

'Is it wise, sir?'

'It is an order.'

Pugh was glaring at his captain, but he did not dare to oppose. Only the Gunner, who could be as fiery as his weapons, took his own stand.

'Sir, a hundred slaves loose on the upper deck is more than I can control.'

'They will not be loose. They will be manacled.'

'They may still make a rush at us, sir.'

Matthew had had enough of this. 'How many muskets and hand-guns have you?'

'Twenty, sir.'

'And the men to use them?'

'Aye.'

'Twenty muskets against a hundred slaves in chains! If you lose that battle then you are a slave yourself! Now fetch your weapons and say no more.'

If Matthew had hoped for some forgiving wind of heaven, or the pure blessing of sunshine, to work a miracle upon his slaves then he was doomed to disappointment. They emerged on deck like the ghosts of manhood, pale wraiths of the living, crucified by men and now mocked by God.

Black faces had turned grey, the colour of rain-water on a dirty street. Bent bodies remained bent, as if invisible cages still confined them. They crawled, they shuffled, they staggered in their desperate weakness. Some dragged bloated corpses in their wake. The few who could crack thir sinews to walk upright gazed about them in hopeless apathy, while those who had never seen the sea stared at it with dull horror, as if it could only be a worse form of torture.

In this man-planted garden of the damned, miracles must be worked by men.

The Gunner and his mates stood bravely to their arms for some few minutes, alert for treachery, then for very shame they shed their martial valour and stood slack-handed, merely foolish. The threat from this throng of tattered crows could scarcely menace a crippled cook. Matthew was the man to feel the worst of guilt. Surveying his shambling flock, his own children, his private graveyard of dead and wounded and doomed, he knew not what he should do to atone.

It was Doctor Rushforth who now forgot his pomopus airs and at last showed his true mettle. On a sudden, he assumed the posture of command.

'Bear a hand here,' he called to Harkness. He had to raise his voice against the background of wailing and groaning, and it seemed to come easy to his spirit. 'First we will douse them down, with a strong firm spout of sea-water. I cannot work in a midden . . . Have the goodness to rig a pump for me, *not* down to the bilges but over the side, so that we suck up something clean. The foulness can run out through the scuppers, and good riddance to it!'

Harkness hesitated, looking towards his captain, while Boatswain Pugh dared to give an exclamation of disgust. White men *working* to cleanse this black filth? They would be serving their breakfast in bed on the morrow! . . . But Matthew only nodded and said: 'Turn to, Mr Harkness. Rig a pump, and whatever else the doctor may call for.'

Thus it was done, and continued to be done, for the best part of a working day and far towards the evening. Matthew was amazed at the infinite care with which Dr Rushforth, a man transformed, went about his healing task. Having disposed of the first crusted filth of his patients, in batches of ten and twenty at a time, he then turned to the gentler arts of medicine.

At the beginning he could not be so gentle: the dead must be unshackled—or, under Pugh's heartless direction, hacked off from their chains with the loss of a foot or a hand—and cast over the side; then it was the turn of the dying, whose unstirring bodies, in which only the tormented eyes remained alive must be judged to be past redemption.

Then the Hippoctratic Oath took command. Rushforth, turning his back on the butchered limbs and mountains of ordure which were the shame of medicine, became Luke the beloved physician, and assumed the role of tireless saint. He purged, calmed, bandaged, staunched, and fed a multitude of unfortunates, and gave them hope, and himself salvation. While the foetid decks below were washed down and sulphured, he made ready as best he could their former inmates for, at the least, a softer spell of purgatory.

He took an old man, with all the shudders of eternity racking his body, and sponged his face, and spooned into a trembling mouth some beef broth which, a little earlier, might have served as the essence of life.

The old man died, but others near him, watching this loving care at work, gained heart from it, and lived to remember it.

He took a woman who had survived a still-born labour, and the puerperal fever which accompanied it, and cooled her fiery body with vinegar, and sutured the wounds to a torn pelvis. She lived to thank him in a tongue unknown, with a trembling smile which needed no translation.

He took a weak, skeletal waif of a child, with the enormous eyes of dumb suffering. This starveling, with the mockery of a swollen stomach to deny its famishment, would not eat or drink. Rushforth, with firm hands, made it submit.

He pinched its nose until the gasping mouth opened, then tipped in some of the broth. The child wrenched its body aside and screamed, and at last swallowed; while some women round about screamed also, and would have rushed to its rescue if the Gunner's men had not forced them back—their greatest battle of the day. Peace at last returned, and with it a small rebirth of life.

He took hundreds of others into his care, making of them what he could—which was sometimes all, and sometimes nothing. He worked for fourteen hours on that day of contrition without respite or sustenance save for generous stoups of wine. Matthew would not have begrudged him a barrel of it . . . When his frock-coat became too foul with blood and filth, Rushforth threw aside this butcher's apron and, stripped like any slave, toiled on.

When at last he was finished, and the decks cleared, he stood up trembling, leaning back against the bulwarks in utter exhaustion; and silence fell upon an evening which might just have redeemed a handful of men—himself the peerless first—before the wrath of God.

In the calm of his cabin, Matthew thanked the surgeon handsomely. For the first time he understood why Benjamin Boothroyd continued to employ such a seeming mountebank. But

Rushforth's reply gave him the strangest turn of the day.

'It was God's work,' he said. He was near to dropping, and—fresh from pain and naked to praise—nearer still to tears. 'None the less, I thank you for your kind words. They come as —they come most movingly.' Then he brushed his hand across his eyes, and stumbled away.

Matthew at last divined what lay in Rushforth's most secret heart. The good doctor wished to be loved, and never was.

In the happy latitude of 40 degrees west of Greenwich, the North-East Trade Winds took command of their progress; and, from that midnight moment on, the *Mount Pleasant* sailed into the ultimate of peace and beauty, cradled in a sailor's paradise of warm airs and steady mileage.

A prim landsman might call it the second leg of a swinish triangle, the base extension of the Round Trip; seamen who savoured their release into profitable idleness could only stand back and say: 'This is why we volunteered.'

While Dr Rushforth still laboured among the slaves, their approach to the enchanted island of Barbados, a palm-green oasis in a pale blue sea, did much to blur the memory of a disgusting voyage. Presently, after seven weeks, they were threading their way through the shallow waters off the island, sighting a path by church-spire small as a needle point, by distant conical hill, by 'markworthy tall tree' which was still markworthy after its first charting had faded a hundred years.

They were seeking, with work-a-day dog-navigation, the one safe entrance through a coral reef; while all around them the Caribbean stretched towards the horizon in fantastic tiers of colour, from light to dark blue and then to the black-green of sunset.

Underneath their keel was clear water, unsullied by the piggery of man or the dross of nature, the very colour of innocence—there could be no transparency like this in the grim northern climate which they had left behind. Their course briefly stirred the coral rock, leaving a milky wake before its myriad particles settled again. Then God-given purity veiled their passing for ever.

On the last evening of the voyage, Rushforth—to the disgust

of some and the approval of his captain, who had seen the doctor's prescriptions stem the tide of mortality below—Rushforth brought all his charges on deck, perhaps for a farewell kiss of freedom. Twilight hid the gross details of their condition; night, under ghostly billowing canvas with the thresh of water beneath their prow as soft as the first breath of sleep, cancelled much and forgave all.

There was left only endless shuffling, and some groaning, and the rasp of inexorable chain. Then, of all things, when their slaves were settled they began to sing.

They sang a song of Africa, sweet as home, sad as exile, piercing more hearts than the singers could even dream of. Its cry rose gently upwards from the foul hull and the crowded deck, through tier upon tier of clean white sails, to reach the upper air. If it ever reached heaven on the wings of its brave endeavour, then it deserved to.

Could it be that tears might rise as well as fall? When mortal music died, a drenched God might still be listening.

On the morrow, within touch of Barbados, when all such wayward fancy had shrivelled down to ship's business and close acounting, Pennyman the trading mate entered his particular empire again. With due formality he gave Matthew Lawe the story of their voyage, in his own fashion.

'Beg to report, sir,' he began, with the assurance of a man from whom all real blessings flowed, and then read from a well-thumbed ledger: 'Slaves embarked at Whydah, 599. Slaves to be landed at Barbados, 515. Losses on cargo, 84, or fourteen *per centum* of the whole. With the *addenda* of eight new-born infants at breast, the total landing will be 523.'

New-born at breast . . . Eight *addenda* . . . It seemed that life might conquer death after all, or at least blunt its cropping scythe; and life, as they rounded up into the rim of their faithful trade wind and dropped anchor in Carlisle Bay—life, for some, now turned wonderful.

5

Pedro Ferreira the trusted factor, whom Mr Boothroyd, in far-off no-nonsense Liverpool, had called Peter, came on board while the *Mount Pleasant* was still settling to her anchor cable.

He arrived in some state, conveyed in one of his own island schooners, of the sort which spent their working lives coasting down those Trade Winds south to Trinidad and north again to Guadeloupe.

But for all their bold swing across a thousand miles of coral sea, they could be managed to a hair's breadth; and this one touched alongside the slave brig with a nudge which would scarcely have cracked an egg.

It was a prelude to a display of competence and industry which gave the lie to an ancient English gibe: 'South of the Channel or West of Cornwall, it is the men who die in labour.' Ferreira was that rare inhabitant of the tropics, one who disdained both the burden of heat and the habit of sloth, and simply bustled to his work. Already he was bustling bravely.

A small lithe lizard of a man, he climbed the ladder laced to the *Mount Pleasant*'s tall side as if he had a fire-work up his tail and a monkey to press it home. Then he showed himself on deck, calm, smiling, and collected, resplendent in a suit of creamy tussore silk and a fine woven hat from Dominica.

Matthew greeted him in the sole calm of the quarter-deck, while all about them was the fervour of ropes a-coiling, sails a-furling, sailors cheering the prospect of freedom, and slaves within the shadow of their journey's end moaning in fear of fresh terrors unknown.

But for some men there were no shadows, and of these elect, Matthew Lawe was one and Pedro Ferreira another.

'I have not come to disturb you, Captain,' Ferreira said, after he had given a flowery salute with his hat and shaken Matthew's hand. 'Merely to present my compliments and place myself at your service.' He had a lilting way with his speech, using words as the generous currency of good manners, good-hearted welcome, good fellowship. 'I hope above all else that you have had a happy voyage with the smallest losses possible, and are now ready to enjoy Barbados, which'—and again he swept off his hat, but more widely, to include the distant view of sunlit hills, fronded palm trees, and the golden beaches of idleness—'lies completely at your disposition. In a word— *bemvindo*!'

'You are very kind,' Matthew answered. It was impossible not to warm to this engaging fellow, who spoke such fluent easy nonsense while his eyes, sharp as scissors, took in everything around him, from the furl of the topsails to the cut of Matthew's own coat. 'Yes, we have made a brisk passage. There was some fever, but my good doctor kept it within bounds. We will land some five hundred and twenty slaves.'

'Excellent! With a loss, perhaps, of twenty per cent?'

'Fourteen.'

'Admirable! . . . I'll not trouble you with business,' Ferreira went on, as if the concerns of his life were poems and pearls of philosophy rather than the vulgar tongue of commerce. 'With your permission, I will speak with Mr Pennyman before I take my leave. But'—his sharp eyes scanned the horizons beyond Carlisle Bay, bare of any substantial ships—'I had word of the *Mersey Blessing* being with you. Is that not so?'

'We were forced to part company,' Matthew answered. 'Captain Downie was challenged and boarded by a British frigate, off the African coast, while I made my escape. I do not know his situation at this moment.'

Ferreira asked an essential question. 'I trust he was not caught with his cargo?'

'No.'

'A most difficult decision,' Ferreira said—and that was all, in a vile cut-throat world as far from the beauty of Barbados as hell from paradise, but perfectly known to such men of the trade. 'Well, Captain, we shall have much to discuss later. Landing your slaves. Cleaning your ship. Loading her with a choice cargo for the homeward voyage. I have certain hogsheads of blended rum, a hundred gallons a-piece, from my own plantation—and fit for a king! If you will do me the honour of dining with me tonight, we may combine business with true pleasure.'

'That will suit me well.'

'No better than myself . . . I will send my boat at five o'clock, and attend you on the quay. It is a drive of a mile or so; but I have a well-sprung whisky and a fine trotting horse.'

'A *whisky*?'

'Ah—the word is new to you? It is a light carriage, two-wheeled, and goes like the wind. It has *yellow* wheels!' For some reason Pedro Ferreira seemed to be entranced with this idea. 'So, there it will be waiting, all ready for you, and we will drive out and dine at our leisure. And our pleasure. What would you say to a dish of succulent spiced ham?'

Matthew, with a sailor's seven-week lust for fresh meat, smiled. 'I would say, "Good evening, Sir Porker".'

'The very last words it will hear. Till tonight, then, and once more, *bemvindo*!'

Matthew was not to forget the splendour of that late afternoon landing in Barbados, his second visit in all the uncountable years. He had come there first as a pirate on the run, with a price put upon his head by at least two sworn enemies, Henry

Morgan the unrelenting tyrant and Simon Montbarre the Exterminator; now, as captain of a trim ship, both of them new to the coast, he was an honoured guest, and happily made to feel so.

Pedro Ferreira met him as he had promised on the quayside of Bridgetown. His greeting was as warm as the summer-scented air, while at his back the yellow-wheeled whisky and the horse dancing and trembling to be off seemed to add their own delighted welcome.

Once they were away, the carriage ran merrily, like an agile bouncing spider, through the straggling township and then by country lanes past cane-fields where the slaves were still bent to their work, and planters' noble houses, and wooden shacks with neat gardens on display, and a parish church or two; while all around them the soft air, the clear light, the gentle smell of a sea which could no longer harm him, brought balm to a sailor's heart.

Ferreira must have saluted, with his whip to a man and a flourish of his hat for a lady, a hundred times during the course of their progress. All was smiles for this smiling traveller.

'I have taken a liberty,' he told Matthew, as he led his guest up the coral steps and through the portico of a most elegant house, glowing pink in the lowering sun, surrounded by bougainvillea, hibiscus, tall palms, and magnolia like pliant ivory. 'The liberty of inviting no one to join us on this first night. We will talk of your own matters, and we will enjoy. After that, I can assure you that our Barbados society will not give you a moment's peace!'

'I am well content with that,' Matthew assured him. He had paused to take a backward glance, from the crown of the hill down to the distant sea; and there, riding to her anchor, was his own *Mount Pleasant*, safe and sound and patently at ease. 'After a long voyage in a crammed ship, the fewer souls about me the better.' He could not forbear to add, as he gazed around him: 'But this is wonderful! Such a view, such refinement, such peace. One might be in heaven already!'

Ferreira smiled at the words. 'We all make our lives. This is mine. For a space, it is yours to share.'

They dined *en prince*, to the haunting music of a guitar ('My garden-boy is a Creole of Brazil'), at a table of polished yellow-wood shaped like a lemon leaf, with silver-ware which might have come from a Spanish galleon. The fare also was princely: a soup of turtle spiced with nutmeg, steaks of fried dolphin, a magnificent ham sugared and adorned with cloves, and gleaming pawpaw with the taste of honey and ginger mixed.

Salt beef, stale bread, the smell of mould and the stink of misery, all were forgotten. Matthew could only smile, and feast, as he stretched his legs in grateful ease under such a table. He could also listen, with all the pleasure of a man who had earned his keep—even this keep.

Such was his mood of content that it did not seem incongruous to have Ferreira spicing his excellent dinner with close questioning on a far-removed topic—the market for slaves.

'The prices at auction are certainly lively,' the man-of-business told his guest. 'I have never known such eager bidding when the quality is good. In islands such as Barbados—and in America also—the fear now is that the new regulations will serve to dry up the supply. The arrest of the *Mersey Blessing*, the loss of the whole cargo, may well be the pattern of the future. So—we buy slaves, at whatever price, so long as slaves can reach these shores. You will find,' Ferreira added, with a glance at Matthew, 'that there is strong feeling hereabouts against your English Navy.'

'It matters very little,' answered Matthew, to whom nothing at that moment mattered at all.

'Even if it is openly expressed to yourself?'

'Even so. I can answer that the Navy, as always, does what is set out in its orders. If the Navy of today has any motto, it is this: "*They* say. *We* do." So, if a man of Barbados is ready to make a quarrel out of that, he will find himself spitting against the wind.'

'Spitting against the wind?' Ferreira repeated doubtfully.

'There is a coarser phrase, but not to be used in such a house as this . . . I mean, that if a man discharges something at me, it will merely fall on his own breeches.'

Ferreira smiled. 'Well, I thought it was worth a warning to

you . . . It would interest me to know what were the market-prices at Whydah.'

'Da Souza was ready enough to lift them sky-high, I can tell you. He made a most determined assault upon me for two pounds a head. We settled at one pound and ten shillings.'

'You did well,' Ferreira said. 'On the African coast they have certainly sniffed the wind of higher prices here . . . Mr Booth-royd will also do well. And you, I hope.'

'And you.'

'We shall see . . . In the meantime, I will land your slaves tomorrow, and house them in my barracoons down the coast at St Lawrence. They will be brought to auction within the week.'

'And after that—for them?'

Ferreira shrugged. 'There are good and bad masters. Those who are cruel or stupid will always ill-treat their stock. But in Barbados they are not well esteemed . . . It is in the hands of God.'

As if to mark a moment of such fateful decision, a single cannonshot boomed out from Bridgetown below, setting a multitude of dogs a-barking. It was repeated nearby, and then, far to the north, a third shot answered.

'What is that?' Matthew asked.

'The slave-guns. It marks eight o'clock of the evening. They are now freed from their work.'

'Till when?'

'Dawn light.'

An hour later, comfortably installed on the coral-stone patio with its rattan hammocks and arch of trellised hibiscus, under the first stars of night, Ferreira moved to his second topic, with zeal undiminished.

'When the slaves are out, we must cleanse your ship.' The glow of his rare Havana cigar, matching Matthew's own, was a friendly beacon in the dusk. 'You would not wish to load my fine cargoes into such a midden . . . Recently we have been us-ing a new method for restoring slave ships to good condition for their return journey. Tell me, Captain, will the *Mount Pleasant* take the ground safely?'

'Aye. On good fine sand, or yielding mud.'

'Sand we have in Barbados . . . The method is this. She will be beached near the end of the ebb tide, and hauled down immediately. When she is careened on one side, the lower ports are opened; half the ship is flooded, and then scoured and pumped out. Then the rising tide—sweet sea-water, all of it —comes flooding in, and carries off every last scrap of filth. On the next tide, we careen her again on the other side, and wash her out once more. In this way——'

Matthew, not liking this talk of flooded ships and rising tides, interrupted him. 'So my ship is to be sunk?'

'I would rather say, she is to be dipped. For two tides only.'

'What if there is a change in the weather? She would not be *dipped*. She would be drowned!'

'Such changes can be long foretold in these waters. The Caribbean is a friend . . . And this plan is a matter of quick action. We have pumps on floating pontoons. We have five hundred buckets. We have slaves! I assure you, Captain Lawe, it is safe and it is speedy. She will dry out in a day, under our blessed sunshine, and you will have a clean-scrubbed ship at the end of it. The tidal sea is the greatest cleanser in the world.'

'But she must be stripped of everything. All furnitures, all remaining stores, all the men's gear. It is a labour of Hercules!'

'Not so, sir. I say again, we have slaves a-plenty, working like locusts! They can bare your holds in half a day. Your officers and men will be quartered ashore. You yourself may stay here, and welcome, or you may prefer lodgings in the town, which I can provide.' A ready smile, with a certain slyness to it, came to Ferreira's lips. 'Are you married, Captain Lawe, may I ask?'

'No.'

'But fond of—of society?'

'Certainly.'

'You need not lack it . . . I had in mind a small dinner-party in your honour, between tomorrow and the auction day. One thing will lead to another, beyond a doubt. Captains who are prepared to run the blockade with such vital cargoes are assured of a particular welcome . . . So——' Like a general giving the summation of his battle plan, Ferreira concluded: 'Let us say, three days to clean your ship and dry her out. Three more to auction-time. For careful loading of your new cargo, perhaps another week. How long do you wish to stay on the coast?'

Matthew looked about him, savouring all the precious wonders of a journey's end: warm and windless air, a star-pricked sky, the fragrance of folding flowers, a house of peace and comfort, the remembrance of a splendid dinner, cigar-smoke, the salt of good fellowship. He could only answer:

'Forever!'

Ferreira laughed aloud: 'Would it could be so. And now, with so much settled, a bowl of my best rum-punch will settle all else.'

Matthew, sitting between Pedro Ferreira and his own Mr Pennyman before the great auction block of Bridgetown, under the shade of a gaily striped canopy, was well content with his circumstance after six fortunate days ashore. All had gone fruitfully: none of his precious slaves had died in the barracoons: Ferreira had guided him in every last degree like a loyal friend; and the cleaning of the *Mount Pleasant* was safely accomplished.

Though it had been a small agony to watch his prostrate ship twice engulfed by the tides, with no effort made to come to her rescue, he had the reward of patience. The *Mount Pleasant* had suffered, and triumphed, and risen again like the phoenix; and the river of filth which had poured out from her hull, to be borne away by the faithful ebb-tide, was the strange proof of salvation.

Best of all, he found that this auction morning, which might crown his first voyage with success, had nothing of guilt or misery about it for the good folk of Barbados, nor for himself. Conscience had melted with the sun; black faces turned shadowy, becoming mere ledger-entries; for the rest, this was a modish, colourful, and festive occasion.

All about him Barbados was *en fête*. His, it seemed, was the most generous cargo to be landed for some months, and after the straitened times dictated by the 'Damned Abolitionists' the island was more than ready to celebrate it. Matthew himself, in a white duck tail-coat trimmed of its badges of naval rank, but with sufficient gold braid remaining to indicate a captain, was the object of attention. It was known who he was; and who he was, was welcome.

The centre of the present scene was the 'block' itself, a huge stone table jutting forward like a playhouse stage, with the auctioneer's dais enthroned on high at its back. On three sides round about it were cushioned benches for the elect of the audience. All had donned their best rainbow finery for this, the strangest of all plays.

Fashionable women in brilliant silks and satins formed the inner ring; they made great stir with fluttering fans, and parasols slyly lifted, and slippers peeping out, and teeth a-gleam with merry laughter, and the chatter of magpies come to a private parliament.

The white men of quality were more solemn, though not less magnificent. Here and there a uniform showed the brave scarlet of the garrison; tall hats gleamed in the sun, frock-coats of pale blue and yellow advertised a thriving tailors' trade; thigh-boots shone as if they were pillars of flame, ruffles and stocks frothed upwards like the snowy bubbles of champagne.

There was much raising of hats and exchange of bows among these gentlemen: bows stiff, bows condescending, bows familiar. They, like their boots, were the pillars of this structure. But they were not the last of it.

At their backs, in the outermost ring of all, was a great concourse, stretching to the green margin of the Garrison savannah, of carriages, light whiskies, two-horsed curricles, family coaches: all attended by the favoured of another race of men —liveried coachmen in top hats, running footmen, grooms, little black 'tigers' who sat on the tail-box behind their masters, their arms folded breast-high, and grinned down at the less fortunate world.

These also had come to see the play. If the actors were their own blood-brothers, they were as yet no more than common slaves awaiting purchase.

Turning his eyes away from such light-minded frolicking, a mere ballet before the solid fare of the drama, Matthew perused the hand-bill distributed to all those nearest to the auction block. It was an elegant manifesto, printed in bold red, white, and blue, adorned with the royal dolphins of Barbados, the curling waves of the ocean, and a representation of a sailor shaking hands with a planter, above the motto: 'Hands Across The Sea!' So far, so good . . . Then he read, with some surprise, the first few lines of announcement:

To be Sold *tale quale* on the Block at Barbados,
In Two Main Lots,
a Sumptuous Cargo of Slaves
in excess of FIVE HUNDRED,
Lately Landed from a Well-Found Ship
By a Bold Captain.

He touched Pedro Ferreira lightly on the arm.

'I have been shown other hand-bills of this sort,' he began, 'by Mr Boothroyd in Liverpool. They set out the name of the ship, and her captain, and the date of her arrival. Is this now changed?'

Ferreira nodded. 'For the sake of prudence. We do not wish to lose you, Captain Lawe.'

'Lose me?'

'By naming you and your ship. All we know, as we sit here at auction, is that five hundred slaves have miraculously risen from the sea and landed on our coast. We need no names. Especially, we do not want some spy, some informer, sending a hand-bill to England, saying "Here is proof!", and giving news of your ship. Naturally, it is widely known. But we prefer to protect you. I may tell you, you and your enterprise are already much admired on the island.'

By chance, Matthew Lawe raised his eyes from the hand-bill at the moment when Ferreira pronounced the word 'admired', and found himself looking directly at another pair of eyes, feminine, frank, and—it might be allowed—admiring also. Seated within ten yards of him, across the angle formed by the front and side benches, was a rarely beautiful young woman who, with no pretence of fluttering fan or dipping parasol, was staring boldly at him.

Matthew bore the inspection as a sailor should, splicing eyes with this paragon in a firm and seamanlike maner. After a long moment he was brave enough to bow towards her, whereupon the lady gravely inclined her head, and with a slight smile turned away.

The deed was done, and, by all the rules of a delicious sport, he was free to examine her at his own sweet will.

What he examined was a sailor's landfall in a dream of desire. He saw a face of creamy loveliness: hair of dark and alluring abundance: shoulders leading gently down to shapes of firm promise—and all this costumed in the most demure elegance as if fashion must deny what womanhood might freely offer.

Of long experience, he knew that this formed part of the best-known female subterfuge. Women—many women— teased and provoked beyond endurance, and then, when presented with the bill of account, denied all knowledge of it. But a thirsty sailor was ever ambitious. He had not come to Barbados to be fobbed off with 'La, sir, you are too bold!' It might well be that she was not in Barbados to cheat an honest man with such fool's gold.

He was determined to find out.

But now he must stop his staring. A slight movement of those delicious shoulders warned him that she was turning towards him again. It was now her moment to examine, and they both knew it. He bent his gaze downwards to the hand-bill as he came under scrutiny.

Suddenly it seemed dry reading, even though it was his sole livelihood.

DAY THE FIRST
Lot One, to be Subdivided *ad libitum* and Sold Without Reserve.
Some TWO HUNDRED AND EIGHTY Slaves of first
quality, likely to make trusty House-servants and Waiters,
stalwart Field Hands, and Labourers.
Among them Thirty *Females,* warranted clean, Persuadable
Maids, Cooks, Dusters, Good Breeders All.
Children under Two Years will form part of the Lot,
for one-tenth purchase price of the Female.
Many Likely Boys. A few Aged and Crippled sold without

Warranty. No Violents, No Runaways.
GOD SAVE THE KING!

Matthew looked up, locked eager eyes with Madame Unknown again, and turned to Pedro Ferreira. He must know the truth of her, and soon.

He opened his mouth to put a formal question, and was astonishingly forestalled.

'I will be happy to tell you,' Ferreira said, solemn and detached as a courier with a foreign traveller in his charge. 'She is Mrs Strutt. Her baptismal name is Mariana.'

Matthew's mouth remained open. He could almost imagine his informant adding: 'The eastern transept is much admired.' But instead Ferreira said:

'She is a widow.'

'Ah.'

'She resides with her father, Mr Gordon Makepeace, and keeps house for him. Mr Makepeace is the foremost of our planters, and his mansion is allowed to be the most beautiful in Barbados.'

'Indeed?'

'Mrs Strutt has recently become engaged to re-marry.'

'Oh.'

'The happy bridegroom-to-be was the colonel of the regiment quartered here. Colonel Sir Thomas Cottle, Baronet.'

'He *was* the colonel, you say?'

He returned home when regimental duty was completed. Mrs Strutt will doubtless follow him there. They are to be married in England.'

'When will that be?'

'In due course . . . If a friend may presume, I wish you good fortune.'

'But she is engaged to marry!'

Pedro Ferreira, who for some reason was now gazing upward at the sky, answered: 'It might be early days to consider it a sacred bond.'

A single summoning bell struck loud and resonant above the auction block. Silence began to fall as Mr Pennyman the trading mate, who had been fathoms deep in notes of account, columns of figures, pieces of paper great and small, joined the living world again.

'Thanks be to God!' he exclaimed, not piously. 'One might think we were here to see the circus!'

The auctioneer, a bustling red-faced man of equal paunch and presence, mounted the steps of the rostrum, bowed to the sea of quality surounding him, raised his silver mallet, and brought it down slap! on the mahogany lectern in front of him.

The sullen bell clanged out once more, and then, to the sound of whips and curses, the first of the slaves began to shuffle out of their stockade and onto the last public stage of their lives.

They had a better presence than when they were off-loaded from the *Mount Pleasant,* better still than on the day they had been dragged up from her slave-holds at sea; but they could not be called a brave company. Though no longer shackled together, they were privately secure: each man's badge of servitude was now a heavy chain some three feet long to which was secured a twenty-pound ball of iron.

This he carried in his hands as he walked, and, when he stood still, must drop on the earth to bring him to anchor.

They circled the block at a limping laggard pace, almost within touching distance of the spectators so that they might be plainly seen. Men and women were clad alike, in a single cotton apron reaching from the waist to the knee; all else was bare and shameless. Children walked naked, clutching their last link with hope, their mother's hand. The shuffling ranks were attended, urged on, held back, pushed into line, by overseers armed with whips, and a regiment of weedy fellows who were the auctioneer's clerks.

These latter were the boldest of the bold, aspiring to manhood with shouts and shoves and continual cries of 'Show your teeth, damn you!' whenever these necessary guarantees of health were hidden. Thus it was that every one of these abject slaves made his tour wearing the most mirthless smile to be seen between the Carib Sea and Africa.

As they disappeared from view behind the block, a torrent

of conversation and comment broke out among the spectators. 'They approve the quality,' reported Ferreira, whose practised ear could detect such tones and tunes. 'I believe this will be a good sale.'

Then the first of the procession reappeared, with a clutch of slaves mounting the block for full and final display. The auctioneer hammered once more, and announced:

'Sub Lot the first: a fine parcel of thirty men. What am I bid?'

'So many at one time?' Matthew murmured to Pennyman.

'At the beginning, sir, yes. These are labourers and field-hands—good health and good muscle.' Pennyman was most knowledgeable. 'The Harbour Master is seeking fifty such for loading and handling. Mr Makepeace of the big house has sold off some of his old stock, and needs at least thirty for his cane-fields. The barracks are being extended, and there is talk of a new fever hospital. That will bring in the Surveyor-General, wanting up to fifty stone-cutters and hauliers. So—the auctioneer suits his tune to the market. After that, the slaves will be sold in twos and threes, or singly.'

'And the women?'

'The same. They go for house-servants and maids. If they are uncouth, for field-hands. If they have looks—for those who like such looks—for what you will!'

'That is permitted?'

'Well, it is understood. They are slaves, when all is said. Who is to tell if a girl is bought for a lady's maid, or to keep a husband in good humour?' The trading mate lowered his voice. 'There is a gentleman here who has spent a small fortune in such purchases. The famous Mr Tinderman of Oistins Beach. Perhaps we shall hear from him today. But we shall not hear from Mr Tinderman himself. *Mrs* Tinderman, by her agent, will bid for a cold-room maid.'

'What is this?' Matthew asked.

'Well, sir, it is a joke on the island.' Pennyman, aware of displeasure, was defensive. 'The auctioneer marks out a likely girl, a mulatto or a griffe.'

'A griffe?'

'A quarter-white. They can be very beautiful . . . When she comes to the block, he will call out, "Here we have a fine strong young female. Would make an excellent cold-room maid." If the agent buys, we all say, "Ah, another cold-room maid for Mrs Tinderman. Her cold-room must be thronged!" But the girl will not be cold for long.'

The bidding began.

While Mr Pennyman, as near to happiness as such a man of strict account might approach, noted down lot-numbers, and figures, and prices, and totals, and subtracted commissions and expenses and docking dues, Matthew listened to the gabble-gabble of commerce, and found his interest in it flagging.

Though it was his own profit, hidden somewhere in this jumbled market-bag, which was at stake, and his first chance of fortune which was being born under the Barbados sun, it seemed far from his true calling, far from a real world.

His thoughts wandered, and they wandered home. That real world was ships and the sea. It was finding a faithful wind, and clinging to it, and wrestling another thirty miles a day from a ship which needed to be coaxed. It was looking upwards at ragged midnight clouds, and finding a break in them, and taking a swift star-sight before the great chart of the sky was blotted out again. It was patience, and endurance, and the blessed dawn to bring comfort.

It was *not* this shameful huckster's cockpit. It was not his forlorn cargo which now shuffled and limped across the coral block of their despair. It was not the sale of flesh, nor compliments on the quality of his merchandise, such as one friendly butcher might pass to another.

Beset by this brief and foolish discontent, he stole a glance at the mysterious Mrs Strutt, to discover if her thoughts inclined towards the business of the auction, the social graces of those about her, the daydreams of womenkind, or to himself. There was no very clear answer.

She seemed engrossed with the gentleman of substance who stood behind her: a large robust man of middle age, red-faced, full of confidence and good humour, attired in the sporting mode like an English squire with money burning the pockets of his elegant breeches.

Was this her father, the redoubtable Mr Makepeace? A suitor of mature years? *A rival?* Matthew could not hazard a guess. All he knew was that Mrs Strutt, lately so forward, was no longer looking at himself.

But soon there came a diversion. The larger parcels of labouring men had all been sold, at fair round prices which caused Pennyman to nod his head with satisfaction. There had been a brisk market for fifty or so of a more intelligent breed, who might be trained to house-service, the management of horses, tavern waiting, even as tally-keepers able to count cane-stalks and bags of sugar.

Now there were women mounting the auction block: women fat and thin, ugly and less ugly, agile and awkward. But among them was one who might be called comely: a girl of fair skin, tall, slim, magnificently breasted, and scornful. She climbed the steps as if she did not care a penny-price for her staring audience, and stood before them like stone—or as a pearl before swine.

Matthew recalled her from the night when the slaves had been brought on deck, though then she had been filthy, and her hair coarsely matted like straw in a midden. Such appetites were not for him, in any case . . . Now he looked at her, as they all looked, and as the auctioneer looked, until the moment was sharpened by a magic phrase.

'A choice lot,' the auctioneer declared, suddenly in league with his audience. 'Mulatto or quadroon. Young and well-fleshed. A likely breeder, or in-door servant.' And then: *'Would make a good cold-room maid.'*

There was titter of merriment among the informed, which presently spread to all the audience. The girl, who could have understood none of this, continued to pose before them, still as a statue, marvellously topped and tailed, but detached from them all, and smokily insolent.

Matthew watched, then—weary of the vulgar byplay—looked away, and met once more the eyes of Mrs Strutt.

She must have known well enough where his attention had been centred a moment before—on the mulatto girl's flowering bosom. She held his gaze for a short while and then, demure as any picture of innocence, dropped her glance to her own pretty endowments. Then her head inclined to one side, and her arched eyebrows were directed at Matthew again.

The message was crystal clear, and might have stunned a mule. It was: 'Would you wish to compare?'

He awoke from a dream of God-knew-what to hear the auctioneer call: 'At thirty-two pounds—once—twice—thrice!' The silver hammer came down with a triumphant ring. 'Sold to Mr Amos Butler, in trust.'

'The very man,' said Mr Pennyman.

All the *Mount Pleasant* cargo of Day the First was sold by half past noon, from the opening lot of sturdy men to the last cripple and suckling child. Mr Pennyman, furiously busy with his reckoning, at length pronounced a most favourable verdict.

'We have never secured such prices before,' he declared. 'They even atone for Mr da Souza's determined greed at Whydah . . .' He consulted his notes. 'Sold, two hundred and eighty slaves, male and female, and four infants. Total amount bid, ten thousand and six hundred pounds. The average price, a few shillings under thirty-eight pounds for each.' He blew out his fat cheeks with something like rapture. 'If we do as well with the second day's sale, we should clear sixteen thousand pounds on the outward passage. With prudent investment here, perhaps thirty-five thousand pounds on the voyage.'

It was a huge figure, and Matthew, basking in the happy sunshine and the golden prospects which seemed to await him, felt splendidly at ease. Round them, all was now in fluttering movement, with the spectators quitting their seats, forming into groups, laughing, chattering, showing their pleasure in a fine morning's entertainment. This was the life for him, if it held free-going seatime as well as profit . . . He spared a thought for one of his smaller burdens.

'We owe much to Dr Rushforth. How does he do?'

'Not well, sir. He cannot shake off the fever. But he is safely berthed in the good sisters' hospital.'

'They will nurse him back to health,' Ferreira declared. 'Prayers and poultices and a woman's care—who could ask for more? And now, Captain Lawe—shall we say, an introduction or two?'

'Certainly. If you think it a suitable moment.'

'I have never known one more so . . . Pray come with me.'

Seen from a closer vantage point, Mr Gordon Makepeace was a most formidable figure of a man, and his daughter, for other reasons, not to be forgotten in a long day's journey between dawn and dusk. It was Mr Makepeace who took the lead, with the utmost affability, as soon as Ferreira had presented Matthew to them.

'Captain Lawe,' he said, 'let me congratulate you on a most successful voyage. You have brought us what we need, and we are much in your debt.'

Matthew bowed. 'It is a pleasure to be here, sir.'

'We must see if we can add to that pleasure . . .' He turned to his daughter, who was standing still and demure, the rim of her parasol just shading her face. 'What think you, Mariana? Should we not do our utmost to reward the captain for his endeavours?'

'I am sure our *utmost* could never match what he deserves, papa.' Her eyes met Matthew's for a flickering moment, but did not hold. With so many other eyes upon them, they were scarcely glancing at each other. It was enough to hear her pronounce the word 'utmost', with a certain intimate inflexion, to be confirmed in hope. 'Perhaps he may join us now,' Mariana Strutt continued. 'I am sure that, after this warm morning, he will welcome some cool comfort.'

'It has proved a warm morning for me also,' her father said with pretended ruefulness. 'I have never known such prices at the block . . . What may we do for you, Captain Lawe? Are you engaged to dine?'

'Why—no, sir.'

'Then pray join us. Mr Ferreira?'

'With great regret, Mr Makepeace. I have pressing business to attend to.'

'Some later time, then. You are always welcome. Mariana, when we can escape this crush, will you conduct Captain Lawe to the carriage? I will follow, as soon as I have signed the bill of sale.'

'Yes, papa.'

They strolled across the green of the savannah, aware of

each other, aware of sunshine and contentment, and, for all the crowds, alone. Her hand rested gently on his arm, with the little finger nestled within that small artery in the crook of the elbow, which only men of science would deny led directly to the heart. At one moment, staring straight ahead of her, she murmured:

'Is it true, Captain Lawe, that your ship is called the *Mount Pleasant*?'

She spoke the name with a curious slowness, dividing the two words as if to give them some secret meaning. It might have been innocent, it could have been lascivious. Matthew could only answer:

'Yes, ma'am. It is a certain hill in Liverpool.'

'Is it, indeed? A pretty mount?'

'Not the prettiest in the world.'

'What should *that* be called? If one ever glimpsed it?'

Matthew could play this little game, and yearned ardently to do so. 'Perhaps, Mount Wonderful? Mount Rapturous?'

'Ah, if one could discover *that*, and climb it!'

'It has been done, ma'am.'

'By you, Captain Lawe?'

'Not in a hundred lonely years!'

6

If Ferreira's hospitality had been princely, this was positively regal. Mr Makepeace's 'great house' was nothing less than that: a colonnaded mansion of virgin pink coral, roofed with the scarlet tiles of Italy, set on the crest of a hill with one view to the rolling Atlantic and another, softer, to the inward arm of the Caribbean Sea.

The outside was a noble mask for what lay within, which proved a dream of spacious silken grace. All the main rooms were oval, as was the shape of the house; from the hall-way, one staircase floated up, as if to cloud-land, and another down to keep the balance of nature, forever grounded in the prosperous earth. But here the prosperous earth was shining pine-wood, without a vulgar nail to be seen, pegged to its base as the stars were pegged to the sky, and polished until it could mirror heaven itself.

102

The house was called 'Prospect', and, for Matthew Lawe, was the finest he had ever seen. He began once more to sink into ease, aided by bright eyes which, glimpsed later across a throng of other people, seemed delighted to have him in company.

'Sit you down!' had been Makepeace's kind command, as soon as the last dog barked, the last dust of the road was flicked from their shoulders and boot-tops, their hands had been rinsed with ewers of lemon-water, and they were bestowed in the inner court-yard. 'Others will be joining soon, but we have time for a quencher before we need stir. What think you of a sangaree?'

'I might think very kindly of it,' Matthew answered, 'if I knew what it was . . . Should I say yes, in any case?'

'Well, it is our favoured mixture hereabouts. Spiced Madeira wine, with certain refinements.' He glanced sideways at his daughter, who was busy with a peacock's-tail of flowers on a side table, and added quietly: 'Good for *all* the parts!'

'It is a strong brew, Captain Lawe,' a cool voice warned him. He did not know how much she had heard. 'Even the name comes from the Spanish word for blood. And Spanish blood runs hot!'

'Thank you for your caution, ma'am. Now I know not what to do.'

'We must put it to the test,' Mr Makepeace decreed, with no further delay for argument. He beckoned to his wine-butler, a magnificent scarlet-sashed figure awaiting any command. 'Joseph! A pitcher of sangaree, with a dish of nutmeg. And for you, Mariana?'

'A cool glass of pressed limes.'

Matthew, already bewitched, chose his own translation: 'I need no sangaree!'

Presently carriages began to roll up to the main portico, and the courtyard filled with the company bidden to dine. There were perhaps thirty of these, and Matthew soon lost the trail of names and faces, full beards and downy mutton-chop whiskers, matrons' bones and maidens' hopeful flesh.

Colonel This and Sir Toby That, Lady Nameless and Miss Bettina Blank, the Bishop of the Windwards and Leewards (could this be true?) and the Chief Justice of—a simple sailor was easily confused, and a sailor with three glasses of sangaree under his belt did not give a finger-snap for any of it.

But whether in the sunshine of the courtyard or the cool shade of the dining-room, where a liveried slave-footman was allotted to each two guests, the principal fare was a succulent pepper-pot stew on which the house of Prospect prided itself, and, in the musicians' gallery above, a trio of pale and seemingly famished Jews played the airs of far-off Vienna, one feeling seemed to unite them all.

This binding element was the cordality with which Matthew Lawe was received.

One guest complimented him on surviving an arduous voyage, another on the quality of his cargo, 'the finest we have seen here for months', and a third on his bravery in running the blockade so that Barbados might be supplied. A stiff remark from a certain choleric planter on the subject of 'the treachery of the Royal Navy' was quickly and adroitly buried by a buzz of further compliments.

Matthew, sitting next to Mariana Strutt at the foot of a noble table, confessed his deep gratitude for all this kindness.

'They mean every word, Captain Lawe,' she told him, 'so the welcome is not forced in the smallest degree.' She turned towards him, as if to press her own welcome, and no man could have denied that she had much with which to press it. 'You must know that you are an undoubted hero in the island. I am proud to have captured you.'

He met her eyes. 'It cannot be your first capture, ma'am.'

'Nor yours, sir.'

The formidable lady of title on Matthew's other side, who had been listening avidly—and was at the age of *nil desperandum*—sought to take her own part in this engagement.

'Pray tell me, Captain Lawe,' she said, 'have you ever seen a sea-serpent?'

'No, ma'am.'

'Or a mermaid?'

'No, ma'am.'

'Or the great Flying Dutchman?'

'Not even he.'

'Have you a wife in every port?'

It must, he decided, be ascribed to the sangaree. 'I am not even married once, ma'am.'

'Oh.'

At his side, Mrs Strutt murmured: 'That does *not* answer the question.'

How could he ever leave this alluring paradise?

Afterwards, with the guests dispersed, Mr Makepeace, who was flushed with wine and already yawning cavernously, said:

'Well, I to my siesta. Pray stay as long as you wish, Captain Lawe. We sup at seven or so . . . Are you living on board your ship?'

'No, sir. There is too much disturbance, and worse to come as we complete our storing. I am lodged in the town.'

'Papa,' said Mariana Strutt, as if struck with a sudden amazing thought, 'might not Captain Lawe lodge with us until his ship is ready?'

Makepeace looked from one to the other before replying: 'I am sure he would be very welcome. He must consult his own convenience, of course. He has a ship in his care.'

'But what would Captain Lawe think if we did not invite him?'

'What would Captain Lawe think if we do?' Makepeace answered with a gust of laughter. He then repeated: 'I to my siesta,' and shambled sleepily away.

Humour, it seemed, was more free in the colonies.

'Papa is so vexing!'

'What troubles you, ma'am?'

'He knows me too well . . . . But who could be truly vexed on such a day as this?'

They walked in the grace of the garden, under trees laced together to form a cool canopy for their delight, past flower-beds of flame and orange, towards cane-fields which stretched to the eye's limit and beyond. Slaves were at work, and did not see them; garden hands, nearer to the family bond of the big house, bobbed and smiled as they passed; all else was dappled shadow slit with gold, deep peace, and the promised verge of happiness.

She was very beautiful, and frank, and free. Matthew had voyaged past all ordered thought. It would happen to him, or it would not, and only a fool would question what lay ahead, or try to bend it one way or the other.

He said: 'This siesta-time is part of your life here?'

'Oh yes. We have a saying, "The planter rules from his hammock", and it is true. All this'—she gestured with a shapely hand towards the distant fields—'is my father's kingdom and he knows every inch of it, and he rules by watching it, as did his grandfather who built it out of trash-land, out of wild-pig country . . . But make no mistake. If aught goes wrong, he sees it instantly, and snaps his fingers to summon the chosen man to set it right, and it *is* set right. That is why he is king, and his overseers are trusted lords in their own realm. That is why Prospect is the crown jewel of Barbados . . . One day it will be mine, and then my sons—*unimpaired*!'

He turned to glance at her. 'You have your course planned?'

Her beautiful face was almost scornful. 'I have everything planned! Women are not fools, though they may choose to take a holiday, to amuse the world of men—and themselves.' Suddenly she was direct, and without pretence. 'You know my situation, Captain Lawe?'

'I believe so, ma'am.'

'From Ferreira?'

'Yes.'

'As a counsellor, you may trust him . . . I plan to marry soon. I plan to marry a baronet. I wish to be Her Ladyship, and I will be so, within a year.'

'I wish you every happiness, ma'am.'

She laughed, becoming young and pretty and warm again. 'You do *not*! You wish God's curse upon Sir Thomas Cottle, and a brisk usurping by yourself. Is that not true?'

Matthew, never having heard such frank words from a woman in all his life, could only match her spirit.

'It is.'

'Well, I must assure you now, there will be no usurping.

Lady Cottle I shall be, and my son when he is born, Sir Gordon of that ilk. But that is not the bill-of-fare of *today*. Today is still holiday, and I have a scene or two to play as Mariana Strutt.'

'With me?'

'By great good fortune to us both, with you. Who else? You arrived at the very moment when such wayward naughtiness became urgent.' She put her hand on his arm, and gave it an ardent clasp. 'Are we at one, my friend?'

'As soon as may be, by God!'

'So you have taken my measure?'

'I beg you to take mine.'

'Do not beg,' she told him. 'Captains do not beg. Even if a door is locked, which mine is not . . . And do not love me. I shall not love you. I shall want, and give, and you shall want, and take. All else is fantasy—fit for silly girls and boys, for lovers in books. I want no book lover! I want a man, a man for this season.'

'When?'

'Soon. But woo me a little of the way. Lazy lovers are the greatest insult of all! I still have the sillines of pride.'

'You have all things.' They were standing very close to each other, while the birds sang. 'A marvellous beauty, and marvellous wisdom. There are moments when you seem so *old*.'

'Captain Lawe! Is this *wooing*?'

'It is part of wooing, which is honesty.'

'True. You are old too, in the same sense. As old as that first Adam, and here we stand in our garden. What shall we plant?'

'All the seed in the world, in the most wild abundance.'

'Ah! . . . *That* is wooing. Say on!'

On this yielding rung of the ladder they waited two days, while the planter ruled his hammock and all the world was virgin bliss (though they teased each other hotly); and then they waited no longer. But even now, the onset of loving had its own strange formality, like some wild fandango preceded by a stately advance-and-retirement set to the music of the drawing-room.

There came a night when Matthew, tossing restlessly in a bed too large and empty for an ambitious man, and too modest for a contented one, was startled and then set aflame by a knock on his door. The night-light, falling on his travelling clock, showed midnight. Nothing stirred in the warm night: not a board creaked; the pale moon creeping through the curtains shone only for lovers. The gentle knocking could only be——

But it was not Mariana Strutt. It was an old black woman, Hephzibah her maid, who presently stood by his bedside and stared down at him, expressionless, the most guarded messenger ever sent by night.

Matthew had sighted Hephzibah many times before, a tall unsmiling woman who, having once been nurse to Mariana's mother, now had one sole charge in the world—the daughter she had also nursed from babyhood. She remained on guard for ever: her enemy was the world of men, and her weapon an all-embracing, watchful jealousy. To see her standing there, seemingly with some secret purpose, could almost persuade Matthew that he was, after all, asleep and dreaming.

Even the message she brought seemed to be wrapped in silence. She bore, and now lowered towards Matthew, a silver salver, and on it a wine-glass—but empty.

'What is it?' Matthew asked, bemused.

She stared at him as if he were stupid, which perhaps he was.

'Come Missy,' was all she said.

'But what is the wine-glass?'

The old woman shrugged. 'Miz' Mariana said to bring it.'

'Empty? Why?'

'Come Missy, find out.'

It was clear that Hephzibah knew all about the wine-glass, and did not approve of it—even hated it. It was clear also that she was not going to give him another word on the subject. He put on his dressing-robe, and, in equal silence, followed a beckoning finger out of his room, down a long faintly-gleaming passageway, and through the door which he knew was the one he desired above all things.

It closed behind him, and he was alone with that desire.

She was soft beyond all dreaming, framed by silken pillows,

calm as a tideless sea, ready as a bird to fly. He sat down on her bed and, finding that he held the empty wine-glass in his hand, set it aside and whispered:

'Thank you.'

She was smiling. 'It is not customary to bring it back.'

'What else is not customary? You must tell me. What is this empty wine-glass?'

Her eyes become mysterious. 'A token. Do you not know it?'

'No.'

'I thought sailors knew all . . . What do they make of an empty wine-glass?'

He considered, aware that she was tenderly mocking him. 'Sometimes they balance them on top of their heads.'

'Do they indeed? For why?'

'To signal that they are dry.'

She nodded gravely. Then she leant aside to her night-table, giving him a glimpse of creamy bosom in the movement, picked up the wine-glass, and with great care put it on her own forehead, where the raven hair met the glowing skin below. 'Do not sailors' girls play the same trick?'

'I know not.'

'You lie!' But on the instant the glass went back on the table, and the single candle was snuffed out. 'Yet I will match it,' she murmured in the darkness. 'I will lie with you.'

He took her in his arms for the first tempestuous moment. Then he whispered: 'What of Hephzibah?'

'She will sleep, or lie awake, outside my door. She is very strict, very proper. She could not bear to have my sleep disturbed.'

'And deaf also?'

'As a stone.'

When he left her at dawn, stepping delicately over the rush mat on which Hephzibah lay—so still that she must have been wide-awake—Matthew was a rare and gallant wreck, and could not have disguised it. He was also still rapt in Mariana's enchantment. None had ever provoked such ravenous pride in his own body, nor shown it in her own.

From the first few moments when, thinking to rouse her

gently, he had caressed the inside of her thighs and found her already moaning for love, and swiftly, wildly inviting it, he had been borne along on successive tides of desire, some quiet as the lapping of a stream, others like whirlpools for strength and swamping appetite.

She could, it seemed, be everything by turns: a veritable dragon of love, a soft and compliant partner, a sleeping innocent, and then a tormented nymph again, who plucked at his body as if she would devour it piece by piece. Her very first words to him had been: 'Would you drive me mad?' (which proved almost true); and her last, hours and aeons later at a particular moment: 'The true captain of Mount Pleasant!'

And this was the girl who was so resolutely engaged to marry! . . .

In the safe silence of his own room he watched the sun rise and the magic island come to wakefulness, while he thought of the marriage again, and could not credit it. But he must couple this disbelief with that strange judgement she had pronounced earlier, which he had not believed either, but which now, for a wandering sailor who took his lover's luck where'er he found it, had certain callous aspects of truth.

She had said: 'Do not love me. I shall not love you. I shall want and give, and you shall want and take . . .' One could grow to respect such cool appraisement.

But he must love her again, in their own fashion, as soon as God gave him strength! He stretched his weary arms, knuckled his eyes, and, as sun-rise gained, bent wool-gathering thoughts towards another fortunate day. What a paradise was this island, and what a haven its newest-charted harbour . . . one thing at least was certain. There would be no more heart-searchings nor nippings of uneasy conscience concerning the horrors of the Round Trip. The Round Trip was this.

There could not be many days left, nor excuses for his ship to linger in Barbados; and the time sped by as if they both moved in a dream, whether they idled in the sumptuous comfort of Prospect, or explored the byways of the island, or watched the cane-cutting and breathed the sickly-sweet smell of burning as the dross of the crop was put to the torch, or lay in love and slept in languor.

Nothing disturbed them; Mr Gordon Makepeace was all hearty hospitality to his guest: the islanders, also hospitable beyond limit, seemed to be conniving at a love affair which could hardly be secret any more.

Then Pedro Ferreira roused Matthew to his duty.

The merchant, wide-awake as the day he first opened his doors for custom, waited upon him at the great house, expressed the hope that he was still 'comfortably lodged', and proceeded at once to business.

'Your ship is loaded, Captain Lawe. Mr Pennyman and I have cast the final accounts, and I think you will be satisfied. Rum, sugar, sweet molasses, tobacco, cotton goods, matting fibres, coarse and fine canvass, spirits of wine—you will find them all there. I have tried to match the cargo which you brought us, and I believe I have succeeded.'

'Excellent,' Matthew said, with the best enthusiasm he could muster. It was not more than an hour since he had risen from a memorable bed, breakfasted on sangaree and flying fish, and bathed in warm water perfumed with essence of pine-bark; he was as far from the world of matting fibre or molasses as he was from China or Peru. 'So we may sail?'

'At your own command.' Ferreira, watching him closely, noting his indolence and the shadows beneath his eyes, made bold to give him a courteous reminder: 'Mr Boothroyd will be impatient to see the results of this first Round Trip.'

'He can scarcely quarrel with that.'

'I have the impression that he carries a strict calendar in his head.'

'It is so beautiful here,' Matthew murmured, half to himself.

'Beyond a doubt. Indeed, if you suited yourself you might stay forever, and many of us would be ready to persuade you. But September is—well, I need not remind you of the Atlantic in September.'

'Chill airs, contrary winds . . .' Matthew could almost feel them, at that moment of indecision. 'A whole world away from Barbados.'

'Quite so. However,' Ferreira went on, at his most accommodating, 'if you feel you might remain a little longer, it might be worth the delay.'

'How so?'

'I am expecting a cargo of fresh limes from Montserrat. We might find room for, say, twenty hog's-heads aboard the *Mount Pleasant*. A great delicacy in your country, I believe, and thus a great price. They would crown a magnificent selection of our West Indies wares.'

'How long must I wait?'

'Two days, and half a day for loading.'

Matthew affected to consider. 'I think it would be worth the delay.'

'I am sure of it.' Ferreira rose, brisk and agile as always, and prepared to say his farewells. 'I will inform Mr Pennyman, and make the arrangements. When would you wish to meet me on board, so I may show you the stowage and agree on the accounts?'

'Tomorrow.'

'Tomorrow it shall be . . .' Ferreira held out his hand. 'Please give my best regards to Mr Makepeace and Mrs Strutt. I am sure they will be pleased with the news.'

'I do not wish to wear out my welcome.'

A tiny flicker of a smile crossed Ferreira's lips. Another man might have ventured directly on a jest suitable to the broader world, but not this paragon of behaviour. Instead he said: 'I am sure there can be no question of wearing out on either side,' bowed, and strode down the noble steps to his carriage.

Now the news must be broken to Mariana.

Foolishly overtaken by a 'surprise' which had long been hovering on the horizon, as plain as a snow-squall in blissful high summer, the mourning lovers met to confer as soon as messages had been exchanged. There was a bower of palm trees at the end of the 'long avenue' which was Prospect's outlet to the southern ocean, and there they hurried, as to a secret hiding-place, in silence and in grief.

But once there, speech came in full flood.

'So soon,' Mariana said, clasping her hands. 'I was near to

112

dying when I read your first note. I had hoped for *weeks*! Why must you sail? Why cannot you send word ahead that you are delayed? Struck by the plague, if you wish! Your cargo will not spoil. Except for the limes, and who is to know of the limes, if you eat them as you go?' Her thoughts were whirling, and the words with them. 'Run upon a rock! Cut down the mast! *Mutiny!* Oh God, I cannot bear it if you go. We have just begun.'

There was no talk now of 'I shall not love you, you will not love me.' She was distraught, and took for granted that he must be the same. Matthew sought to calm her, out of a heart as downcast as her own.

'I *must* go,' he told her. 'I have my orders—"Sail when loaded"—and already they are stretched to the limit. It is my work,' he insisted, seeing her pretty face on the verge of growing mutinous. 'I cannot turn my back on it. We knew, did we not, that there was always a term to this?'

'A long term,' she said fiercely. 'Would you bed me for two nights, like some wretched drab in the streets, and then walk away, whistling "Rule Britannia"?'

'Ten nights,' he said. 'And each more wonderful——'

'So ten is enough? The great rake-hell is shotten already, like a damned herring? Oh God!' she burst out again, in a different mood. 'Forgive me! I am not angry. I am desolate. You *must* stay!'

'I must sail.'

She spoke without thought, like an anguished child. 'Then I will come with you.'

'What? How can that be?'

She was suddenly all spirit and sunshine again. 'Of course. Why did I not think of it straightway? I will sail with you to England.'

'On what pretext?'

'The best in the world. To be married!'

It was a delicate moment. Was this a proposal? A wild change of heart? Since there was no gallant way of putting the question, he temporized:

'That will need very careful consideration.'

'Oh pooh! I have been *considering* it for six months and more. I can keep him waiting no longer.'

So it was still the Colonel, after all. Matthew did not know whether to be relieved or mortified. Now it could be his turn to exclaim, in a huff: 'Would you have me for two nights?' and totter away into the shadows . . . But there were vast problems concealed in what she suggested, in the sudden flight to England: a fiery scandal, the ruin of her reputation, a husband-to-be who, after such brazen conduct, could not be goaded to the altar with one of his own cannon. Matthew decided on a practical objection.

'We are not fitted for carrying passengers. There is not even a spare cabin.'

'Build one! A hutch on deck, and a mat for Hephzibah—that is all we need.'

'There will be talk,' he said lamely.

'There will be action too! I will not waste much time in my hutch . . . How long the voyage?'

'Six weeks. Perhaps two months.'

She sighed. 'Fifty nights. What heaven! I may even be reconciled to marriage at the end.'

He decided, not for the first time, that he would never understand her. But already he was warming to the new future, in a way now familiar.

'What will your father say?'

'Oh, he will rage a little. Then he will be pleased that the marriage plans are advanced, and he will relent. Then he will even recall that this scheme was of his own invention, and be proud of it. As for the gossips—when I am the baronet's lady, they will take mighty great care that I do not hear what they say. To miss Lady Cottle's ball on her return to Prospect would be the death sentence.'

Matthew stood up, surfeited with words, shamelessly eager for a lover's cure. 'Well, you are not Lady Cottle yet. May I ask, what lies beyond these palm trees?'

'A sort of jungle of bushes and grass.'

'But soft?'

'Very soft.'

'And private?'

'Very private.'

He held out his arm. 'May I have the honour, ma'am?'

'Yes, sir.'

So I am to build a deck-hutch aboard my ship, he thought in something like irritation, as he took carriage next morning for Bridgetown and the anchorage where that ship lay: a hutch for little pussy, whose coat is so warm . . . It was not the most difficult of the contrivances and subterfuges to which Mariana Strutt had brought him, during the past stormy days. But it was the most strange. Perhaps his ship would come to be called the Mount Hump-back.

He foresaw difficulties with a carpenter over-driven, a crew grumbling as to how they could work the sheets and braces with a *village* on the upper deck, a first mate scandalized or lewdly amused. But once again, at a price, troubles disappeared like dreams at waking.

It was Harkness the mate who met him at the top of the ladder with a downcast face and a voice to match. The little man came limping forward to salute, and then said:

'There is bad news, captain. Dr Rushforth.'

'He is worse?'

'Died last night, sir. The fever took him.'

So man proposed, and God disposed, sometimes in favour of man. Poor Rushforth, a martyr to his trade and to the noble compassion he had shown on their recent voyage . . . Poor Rushforth, surely loved at last—and in the very bosom of Christ . . . But poor Rushforth, dead, meant a vacant cabin for his passenger, adjacent to his own.

Harkness was waiting.

'That is a great misfortune,' Matthew said.

'Aye, sir. Burial at noon today.'

'So soon.'

'Tropic regulations, sir. Mr Ferreira has made all the arrangements.'

'Very well. Save for ship-keepers, all the crew should come ashore to attend. Black neckerchiefs.'

'Aye, sir.'

All was too easy. Even the final step, it seemed, was no more than a shallow pace downwards.

That evening, before supper, it was Mr Gordon Makepeace's turn to summon his guest to conference. Pacing the garden below the patio, the planter came straight to business. Mariana had spoken to him, he announced. The voyage to England was a mad scheme, but no more mad than many others she had proposed—and undertaken. None the less . . .

'I will be blunt with you, Captain Lawe,' Makepeace declared. 'You know well enough that Mariana is engaged to marry Sir John Cottle. Between ourselves, he is a dry stick, but the marriage has great advantages, both for myself and for her, and she is quite firm in her decision. I would not wish anything to go amiss with our plans. Indeed, I would go to very considerable lengths to safeguard them. You understand?'

'Yes.'

'Then I may trust you?'

'I——'

'In this island, we speak our exact minds. I should have said, I may trust you in regard to her plans to remarry.'

'Yes, indeed.'

'As concerns the rest, she is young, a widow; very beautiful —or so I am assured—and the marriage day is not yet set.'

'Quite so.'

'Then she may sail with you.' Suddenly Mr Makepeace broke into a roar of laughter. 'You damned rogue!'

Minds were spoken, and humour was certainly more free, in the colonial tropics.

The last of the cargo, the precious limes, had been safely loaded, though not without some danger to life and limb. Since the *Mount Pleasant* was now too deep to come alongside the quay, she lay to her anchor about a half-mile off; and the giant hog's-heads must be rolled into the surf, forced upwards into the waiting drogers by means of a ramp of planks, rowed out to the ship, and painfully hoisted on board.

But it had been done, and now, to the small company on

board—Mr Makepeace, a weeping aunt, some favoured guests full of the spice of this farewell, and Pedro Ferreira—Matthew announced:

'To my great regret, I must up anchor and away. I thank each one of you, from the bottom of my heart, for your welcome. I shall never forget it. Till we meet again!'

There were murmurs from his listeners, then a general movement to comply. The women surrounded Mariana, embracing her; Mr Makepeace gave him a manly handshake and, out of sight, a wink. The listening crew talked among themselves, with many a sidelong glance towards their new passenger. Though a woman on board must bring ill-luck, it was generally held that this would not strike their captain.

Apart from all stood Hephzibah, tall and gaunt and most baleful in her survey of the enemy; and Mr Pennyman, busy to the last with his accounts and his reckonings. Of Mrs Strutt's unexpected presence on the coming voyage, he had only said: 'I will need to know the fare paid.'

Ferreira was the last to bid Matthew good-bye. They stood alone at the edge of the quarter-deck, under a brave display of tossing, flaunting flags and furled canvas eager to be loosed. For the first time, the lively merchant seemed hesitant as he surprised Matthew with a question:

'Before I take leave, may I speak as a friend?'

'You have earned far more than that,' Matthew assured him warmly. To stand on his deck again, at the threshold of the sea, with such a precious cargo of his own to enjoy, had put him in the highest spirits. 'I can never repay you for your help. Say what you wish.'

'Well, it is this.' As was his wont when he had something unusual to impart, Ferreira addressed an object in the upper air. This time it was the distant fore-topmast, with its sails trimmed aback to keep the ship steady. 'There is common talk,' he went on, 'that the colonel, when a young man, had a fierce reputation as a duellist. Even now, he is prone to quite unreasonable jealousies and rages.'

'So?'

'I wish you well,' Ferreira ended, as if his speech had been much longer. 'I wish you *very* well. So I would leave you with some advice given to Napoleon by one of his marshals—who must have been a very brave man, like myself. The nearer to England, the more prudent you should be.'

7

Eastwards they went to clear the islands and their constant traffic and the night-fishing boats which disdained any show of lights as mere cowardice; then northward to catch the faithful westerlies and the gentle swim of the Gulf Stream; then due east again for a straight run home.

It proved a magical journey, with wind and sea as allies to lovers as well as ships: near four thousand and five hundred miles of steady threshing progress, which turned men into contented idlers, and their vessel to a dauntless homing bird.

For the lovers it was all they had dreamed of: long sunny days on deck, when the only thing that stirred was the broad back of the helmsman, and the bustle as the watches were changed: long nights when love flowered afresh, like some rose of eternity. No eyes took note of them; no ghosts from the cruel outward voyage demanded vengeance, or even memory; the sounds of the sea masked the most tempestuous love-making, and lulled them to sleep as if in a cradle.

Old Hephzibah, having routed all adversaries, was installed as cook, cup-bearer, bed-maker, and sentry, forever standing guard in an angle between their two cabins. All was discreetly done, though a blind child would have known their situation. Only Harkness, the strangest messenger of these gods, serving as link between captain and ship, was in evidence from time to time, with reports of their progress and new orders for its continuance.

It would not be long before the mate was ready for his own command, Matthew thought by way of excuse for this loose rein. A few more voyages would fit him for anything . . . But there could be none like this one, none with so rare a delight in every part of it.

Soon, too soon, it drew to its close. The *Mount Pleasant* passed Land's End in forty-eight days and, with even the tumbled Irish Sea proving a quiet friend, four more brought them to their last evening on passage, skirting the lighthouse off Holyhead and hearing the wheeling seagulls cry their welcome.

All evening, Mariana had been tender and companionable —and unaltered. The bride still hastened to the church door . . . But as she stood up, preparing to go below for their last shared night, she seemed to bear the new-born moon on her shoulder, and all love in her eyes.

'Do not delay,' she murmured. 'I will not.'

'I must stay half an hour,' he answered. 'Soon we alter our course. For the last time.'

'Then how long to Liverpool?'

'Perhaps fourteen hours.'

'The dark ones still belong to us . . . Are you man-alive?'

'I will be.'

'I feel that promise already. Oh villain and lover!—where shall I find the likes of you again?'

It was the first mourning cry she had ever uttered to him, the first trembling in the curtain of her resolve. And such a night to choose . . . But he did not believe that, having come so far, having dared so many things—for him, for her own last hot season of love—she would weaken.

Mariana proved him right. She was strong in her desires, wild in their execution, passionate under the whip of love, generous to ensure that he lacked nothing when all the precious flowers were picked. But towards dawn, cast ashore on the soft beach of exhaustion, she was still Mrs Strutt, firmly engaged to marry, constant only in this one resolve.

Matthew had asked: 'Shall we meet again, Mariana? When I return to Barbados?'

'I think not.' She lay on her side, her cheek cradled in one arm, the other nestling on his shoulder. Naked as the day, she was fully armoured still. 'In society, of course. Perhaps you will come to dine. But no secret glances . . . I shall not betray him, Matthew.'

'Not like this, for a season?'

'Not for a kiss in the moonlight . . . It is written in my book, and the book is sworn to. Accept the honour of his proposal. Say goodbye, with a great flourish, to loving like this. Marry him, and honour the bond. *Finis.*'

'But where's the benefit? If you are unhappy——'

'Why should I be unhappy? What man of magic do you believe you are?' But the chiding was only gentle. 'Oh, I shall not be happy like this. That is why I seized the moment . . . But there are benefits to be had, and I shall seize them too. My father is a man of business, a *great* man of business, but it is not rank. My mother was a church missionary, come out to convert the heathen. She met her match in this when she met my father, and loved him till the day she died. But that is not rank, either. *Rank I will have!* I have told you—love is a tune for children to play. Tomorrow I join the great world of the grown-ups.'

'Today. That faint light is the dawn.'

'Today, then, if you must join the great world of chronometers . . . How shall we end it, Matthew?'

'In sleep, I must confess.'

'I am proud of that . . . Forget me not. Pursue me not. We have lived in Arcady. Now we leave, and lock the gates behind us. But I tell you this. If I weep at any time hereafter, it will be for you.'

In a courteous note brought on board in the pilot-cutter, which had been signalled by the coastguard as the *Mount Pleasant* approached Holyhead, Mr Benjamin Boothroyd regretted his inability to meet the ship on arrival, since he had pressing business in Manchester. Would Captain Lawe kindly assemble all papers of record relating to the voyage, and send them to his house by hand of Mr Pennyman? Mr Pennyman would remain there to discuss and inform if necessary, and Mr Boothroyd would welcome his captain in a few days.

Meanwhile, his ship was under bond in Salthouse Dock, and scrupulous care must be taken that this was not breached, since some delicate negotiations were under way.

Matthew, not without misgivings that Mr Pennyman might have a fine yarn to spin concerning the return journey, made these arrangements. 'Delicate negotiations' might be presumed to concern the bribery of that particular official who would set the customs-rate. Then he waited in his cabin, just as Mrs Strutt, having despatched an urgent messenger to Sir John Cottle 'at his estates in Cheshire', waited in hers.

The ship was busy with sailor's business: striking down top-masts and yards, rigging slings for the cargo when it might be released, painting, scrubbing, and stowing sails. But it was not all work. Families who knew within minutes when a hoped-for ship was safely in, crowded the quayside to greet her. Girls who lacked this family tie were equally prompt. Liquor-sellers plied a covert trade: creditors from the outward voyage came pressing in.

Even hopeful land-sharks promised 'a snug berth in a brand-new ship, sailing in four days with cash advances of the most noble generosity'.

After seven months, it was good to be home.

Sir John Cottle, when he arrived in some flurry and state, borne in a crested family coach with out-riders cracking whips and spotted dogs bounding alongside, made his presence felt immediately. He was a small bone-dry man, straight of back, haughty of face: his reddish ferret's face came to a point, which was his nose, as if the whole had been drawn by a carto-grapher rather than a more merciful God.

He was in full regimentals, and his sword tapped against the staves of the gangway like a drum-stick beating the Advance.

Matthew, warned by Harkness, was out on deck to meet him. He saluted most formally, and said:

'Good evening, Sir John. Welcome aboard.'

Sir John, whose stiff-necked insolence was no temporary mask, stared at him for some moments before answering:

'Haugh!'

Already Matthew feared the worst.

'Haugh'—a kind of trumpeting neigh which sufficed for horses, dogs, junior officers, and all lesser mortals down to ship's captains—was, it seemed, Sir John Cottle's favourite mode of address.

'I should have been informed earlier,' he rasped out, with-out preamble. 'What is your explanation? Haugh?'

'There was no opportunity of sending word from Barbados,' Matthew answered civilly. 'We sailed at short notice. No ship except mine could have reached Liverpool earlier.'

'Short notice?' Cottle repeated disagreeably. 'Is that the fashion in which you treat your passengers?'

'It was Mrs Strutt's decision to join us, just as we sailed.'

'*Haugh!*' The Colonel glared at Matthew as if he had been guilty of gross impropriety in venturing the name. 'We shall not discuss Mrs Strutt.' He then proceeded to do so. 'Has she brought some gentlewoman with her, to act as chaperone?'

'Her maid, sir.' The 'sir' was prompted by a pressing sense of guilt. 'Hephzibah.'

'I do not mean a blackamoor,' Cottle snapped. 'A woman of quality, naturally. A relative.'

'No.'

'Monstrous! What was her father thinking of? Where is she to be found?'

'In her cabin. Packing, I would suppose. May I conduct you to her?'

'We will not discuss a lady's *packing*,' said Cottle, who seemed to have an extraordinary notion of the proprieties. 'Yes, take me there immediately. The sooner she is free of these surroundings the better.'

'This way, if you please.'

'*Haugh!*'

At the door of the cabin, Matthew made as if to knock.

'*I* will knock,' said Sir John Cottle.

Matthew left him at the doorway, glad of a time of truce. But it was not more than two minutes before the Colonel came storming up again. He was in a towering rage, and his yellow face was patched with purple.

'This is beyond belief!' he began. 'What do you mean by your disgraceful behaviour? By God, sir, *I have a mind to call you out*!'

'I do not understand.'

'Do not be insolent,' the Colonel roared. 'You understand full well, if you are a mere fraction of a gentleman. *That was Mrs Strutt's bedroom!* There was a bed in it. I saw it. She was —oh—recumbent on it. How dare you take me there? Have you lost your senses? *Haugh?*'

'I thought, in the circumstances——'

'What circumstances are these? What are you hinting at?' Cottle's hand was on his sword-hilt, and both were shaking in communicated rage. 'I will tell you. You are hinting that because Mrs Strutt is engaged to marry me, she may receive me in her bedroom! Is this a ship or a farmyard? Do you hold her reputation so cheap? Is that it? Say the word, and you and I will have a meeting tomorrow!'

'I thought you would wish to be private.'

'*Private?* In her *bedroom*?'

'I beg pardon if I have transgressed.'

'It was an abominable insult. However, if you are prepared to ascribe it to stupidity rather than loose thinking, then there is no more to be said.' But there was a great deal more, and Sir John Cottle would say it, and forget, if need be, every mad taboo in order to make it tell. 'I will touch on one other matter, for your future information—or should I say education. The merest hint is all I shall allow, but sufficient I hope for the thickest skull . . . Though Mrs Strutt has led a sheltered life, she has had a monstrous amount to bear with. That despicable hound Strutt whom she married! She will say little about it, even to me, but I understand that he was—*most demanding.*' Cottle's eyes glittered. 'A woman of her refinement! *Bestial!* If there is any way I can protect her from further degradation, I shall take that way. You understand me?'

'Yes.'

'And in future, never—never—conduct a male visitor to a lady's bedroom. Not if he is the husband. Not if he is the father! Sodom and Gomorrah were founded on such . . . Now, send one of your manservants to instruct that woman Hephzibah to ask her mistress to wait upon me, on deck, at her convenience. It that clear?'

'Yes, sir.'

'I should think so! *Haugh!*'

'Do you not agree, Captain Lawe,' Mrs Boothroyd asked, when the Madeira and seed-cake had been apportioned, the four children had returned to their station on the sofa, and Matthew was uneasily balanced in his armchair, 'that little Dulcie has made the most prodigious progress during the time you have been away?'

Matthew, under the burden of being unable to remember which of the two female children was so named, nodded sagely in their direction.

'Indeed, ma'am,' he answered, 'a vast progress. One might think her a different child.'

Mrs Boothroyd beamed comfortably. 'There, Mr Boothroyd,' she challenged her husband, 'did I not tell you? Everyone remarks on it. She has grown more than an inch in height, she is more robust altogether, and she goes to her own dancing-class. Madame Clegg thinks the world of her!'

'And sends in a bill according,' Boothroyd answered grumpily. He was in a mood of silence, and his greeting to Matthew had been reserved, almost wary. 'Dulcie can grow an inch without any aid from Madame Clegg.'

'Why, Mr Boothroyd! Would you grudge your own child her dancing lessons?'

Matthew, seeing that the answer to this was likely to be 'Yes,' intervened.

'They all seem greatly advanced, ma'am. It must give you the highest satisfaction.'

'Satisfaction *without end*,' Mrs Boothroyd said, and embarked on a discourse to match it. 'Harry is in his first term at school now. He is making *great* improvement. Little George can tell the time by the clock. Effie won a prize for drawing—such a pretty piece, I will show it to you, the outside of this house with a tree and a bird sitting on a bough, I was quite affected. Who knows, perhaps she will grow up to be an artist —won't you, my pet?'

'I want to have a baby to play with,' said Effie.

'Sit up, dear . . . George, offer Captain Lawe another glass of wine. Harry, have you nothing to say to the Captain? Tell him what you learned at school today.'

'Nothing,' said the elder son and heir.

'Fine school,' his father said gruffly. 'You do not pay attention, that's what the trouble is. What are two times five?'

Harry's fingers performed a covert dance on the sofa-edge, 'Ten.'

'There,' his mother said fondly. 'What do you think of that?'

'Bless my soul,' Matthew said. 'I thought it was fifty-five.'

Harry looked at him with scorn. 'Then you were wrong, and I was right.'

'Harry,' his father said instantly. 'Make your apology to Captain Lawe.'

'But he said fifty-five, papa.'

'No matter what he said, I don't care to hear you speak like that. Come, apologize.'

Harry disputed the justice of this, turned sulky, and was sent upstairs. George, a loyal supporter, pulled such a variety of ferocious faces that he was speedily despatched in his brother's wake. This evidence of domestic strife was too much for little Dulcie, who dissolved into inconsolable sobs and was led away by a nursemaid. Effie, a lone survivor with no particular role to play, announced that she was feeling sick, and seemed ready to prove it.

Supper was a meal so silent and thoughtful that Captain Lawe was forced to conclude that all blame was being ascribed to himself.

But once installed with his guest in the panelled smoking-room Ben Boothroyd wasted no time on his errant children. Family life, with all its storms and stresses, was forgotten as the ship-master turned his close attention to Voyage Number One of the brig *Mount Pleasant*.

'It was a fair trip, Captain, a very fair trip,' he told Matthew, 'save for one or two matters I shall discuss later. I have been through the accounts with Pennyman, correcting certain *errors* on the way——' He pronounced the word 'errors' with sardonic satisfaction. 'The provision prices at Kinsale were greedy, but what can one do with those infernal bog-trotters —and what can one do with Kinsale as a loading-port?—our last purchasing for the voyage *must* be made there, and they know it. The slave prices at Whydah were fair enough, and I hope they will remain so. Auction at Barbados, very good, and return cargo provided by Peter Ferreira, the same. I hope you found him of service.'

'Of great service.'

'That man has never failed me in anything . . . So I have passed the accounts. Indeed, your cargo has already been sold, in one parcel, at thirty-four thousand pounds, all duty paid.'

Matthew was greatly surprised. 'So quickly? I thought it would go to auction.'

'We do not sit and dream in Liverpool,' Boothroyd said curtly. 'I had a price offered to me, and I took it. That is what "reputation" means . . . The only delay was with our great Collector of Customs. It needed some determination, and—ah—persuasion on my part, before he would accept my estimate of sea-spoilage.'

'Sea-spoilage?' Matthew echoed the phrase. 'There was none, to my knowledge.'

'Not so. There was thirty per cent, which was allowed as "Condemned—no duty".'

It could not possibly be true, and had the smell of an arrant swindle. But it was not Matthew's concern—unless he were blamed for it.

'No fault of yours,' Boothroyd assured him, catching his thought. 'New ship, new caulking, new working of timbers—together it means a good many leaks here and there. My friend at the Customs House is an old sailing man. He knows this, and accepts it as a matter of principle. So'—the owner consulted a page of figures on his desk—'total profit on the voyage—slave outwards, general cargo inwards—with all expenses paid, and all customs dues levied, amounts to twenty-eight thousand. Your share at three per cent is eight hundred and forty pounds. You will not quarrel with that figure, I dare say.'

There were many answers to this, but—once again—it could not fairly be called Matthew's own business, if he did not make it so. Already he was taking Boothroyd's own infection, that blinded eye which allowed him to see the death of slaves in misery as a mere trading loss, without any other reproach. All that Matthew now saw, from the slippery slope of 'sea-spoilage' and God knew what besides, was a figure, eight hundred and forty pounds—the largest of his life.

'No, I do not quarrel.'

'A glass of wine on it, then.'

He poured the Madeira, and the two partners drank. But the man whose profit-share was some twenty-seven thousand pounds was still dissatisfied, and would not forgo the chance of expressing it. Matthew had met such men before, as captains who thought that commendation was a weakening agent, like too much honey to a bee, and would never allow three words of praise without milking one of them away again. He did not grudge; but he certainly measured . . . He measured now, as the Great Apothecary in the sky, his dose of disapproval.

'How long did you linger in Barbados?'

Without doubt Boothroyd knew, to the nearest hour. 'Three weeks less a day.'

'You took your leisure!'

'With the auction, the ship-cleaning, and the loading, it was not unreasonable. Ferreira's cargo was most carefully selected.'

'Time is money,' Boothroyd said disagreeably. 'An idle ship is like a lazy wife—a drain on the purse, and no more. I was not looking beyond a stay of two weeks.'

Matthew said nothing. Let him look . . .

'I supose you were entertained largely?'

'Aye. They were very glad to see the ship there. I cannot tell you how many compliments were passed on your enterprise.'

'They starve without slaves. I sold them a ship-load. It is as simple as that . . . And you carried a passenger home, I see from the accounts.'

'Aye. There was Rushforth's empty cabin to fill.'

'I do not maintain a passenger service. Passengers *eat*, and lie six feet long for twelve hours a day. The cabin should have been filled with baled cargo, or casks. There was money lost there.'

'But good-will earned.'

'From a female?'

'And her father, and family, and friends. The lady was returning to be married in England. It was quicker to take a passenger than wait for cargo, and load it.'

'H'm. Returning to be married, eh? And the fortunate swain has now claimed her?'

'Aye.'

'What class of fellow enjoyed this special favour?'

'Colonel Sir Thomas Cottle, baronet.'

'*What?*' Instantly Boothroyd was spluttering, as Matthew had guessed he might. 'Cottle of Broomedge?'

'The same.'

'Did he come aboard the *Mount Pleasant*?'

'Certainly.'

'You should have warned me. He is a great man hereabouts. And a very warm man. You should have sent me a message.'

'You were in Manchester.'

'None the less . . . He is a man I would have wished to meet.'

'Well, you have his good-will already. He will be returning to Barbados to assume the management of his wife's estate. That is, when her father retires from it.'

'What man is he?'

'Mr Gordon Makepeace.'

'Bless my soul!' Earlier disapproval struggled with a promise of great future profit in that particular soul, and melted in the contest. 'Well, we'll say no more about that. It seems you choose well . . . If you can warrant me more passengers such as Sir John's future bride, then perhaps we should keep an empty cabin!' Boothroyd shut his ledger, and sat back, closing one voyage, embarking on another. 'You can be off again in three weeks if all goes well. This time, sailing independent.'

'Where is Captain Downie?'

'In Barbados, I hope. *And* all his slaves. He fooled the frigate, which had to let him go. He was back within the month, for a fresh cargo of slave goods. It takes more than a brush with the Navy to stop Downie!'

'He used a foul trick,' Matthew said, ready to speak his mind.

Boothroyd gave him a frosty stare. 'Nonsense! It was quick thinking, and that's what I admire. He lost his slaves, but he kept his ship.'

'No matter. It was plain murder.'

'He is welcome to that . . . Do you know the new penalty for being caught with a slave cargo? One hundred pounds for every slave found on board, *and* forfeiture of the ship. Robbery! Piracy! Is he to submit to that, for want of a drowned wretch or two?' With a full return of his ill-will, Benjamin Boothroyd was glaring at his laggard captain. 'I want more such men, not fewer. I warn you, Lawe. Do not turn soft in this matter, or you and I will fall out.'

8

They did not fall out for six arduous years, during which Matthew made nine full voyages, and the *Mount Pleasant* grew old in service and in sin. Swiftly and sadly she declined from the brave craft she once had been; the Atlantic weather, the extremes of heat on the West African coast, and the constant violence of setting up slave decks and then ripping them out again, turned her, before her time, into a shabby workhorse, hard-driven, abused, falling from grace.

The permanent slave-stink, which could neither be washed off nor masked by sweeter cargoes, now settled like a curse and betrayed to the world her gross employment.

While Barbados continued as the Mecca of the Round Trip, and remained as fair a jewel as on that first voyage, the private magic was gone. Lady Cottle, as she had promised, was transformed into a pillar of social and moral rectitude: one luncheon of starched ceremony on each visit was as much as Matthew could expect; she had her first-born son, then another, then a daughter, with never a hint, never a flicker of an eye-lash, to recall the wilder shore of love.

'One might think,' Pedro Ferreira said, in a rare moment of malice, 'that the children had been brought by the doctor, pretending them to be bread poultices.'

It was conceded that Mariana Cottle, embracing virtue like a diet of suet-pudding at a convent, was lost to the world of warmth and laughter; and since Sir Thomas Cottle himself had now grown arrogant and haughty beyond any dreams of kingship, embalmed in rank like an emperor in a wax-work, the great house of Prospect began to languish in its own musty formality, a place to be avoided if there were any other hospitality in the West Indies on offer.

One other small ghost of the past remained to trouble the

captain of the *Mount Pleasant*. Wandering the streets of Bridgetown while he waited for his ship to be made ready, Matthew sometimes passed the newly-placed statue of Lord Nelson near the Careenage. Though the sun warmed the stone, it failed to soften a former friend. His Lordship looked down with disfavour on this sea-officer who had embraced forever a dishonourable trade. The woman he would have forgiven; the tarnished uniform, never.

Ferreira, who remained a stalwart ally, once presumed on their friendship to question Matthew on the value of that trade, and his prospects in it.

'You are ill-paid,' he pronounced, when Matthew had given him a picture of the ups and downs of sea commerce, the 'bad voyage' which, by cold calculation, consigned more slaves to the deep than to the auction block, and the small share allotted to a captain who took every risk under the sun to achieve it. 'All the human hazards are yours, and yet you advance like a tortoise, when all is added up. You must have a better share in this, or you must turn ship-owner on your own account. Otherwise you grow into a slave yourself.'

It had set Matthew thinking, and added to his discontent. Then a fearsome thing happened which, for this forced witness of it, changed all.

It concerned, once again, Captain Downie—but Downie in deep distress, a tolerable enough occasion for Matthew, save for the path it led him to take. Though the two men had met many times over the years, in Liverpool and Barbados and sometimes on the Slave Coast itself, their relations had never prospered. Downie had grown more capricous, and quarrel-some, and surly with drink; he did not scruple either to cheat Matthew if he could, in the choice of slaves, or to inform against him to Benjamin Boothroyd over some fancied sharp practice which had caused him, Downie the honest innocent, disastrous damage to purse or repute.

Harkness, who still remained mate to the *Mount Pleasant*— 'He is not a captain,' Boothroyd once said flatly, leaving Matthew forced to agree—Harkness had summed it up by chance in the very same words:

'Downie is not a captain,' he had declared, with the freedom of sunset talk in lazy weather at sea. 'He is rogue-meat, clear through to the backbone. He can only grow worse, like from little thief to big thief, saucy dolly to brassy strumpet. Bad ale rots the barrel.'

It was not the mate's place to pass such a judgement on the captain of another ship, nor Matthew's to rebuke him. Let Downie catch his own slanderers . . . Now, when the two met again for the last time, Downie had indeed 'grown worse' in another way—the final one of all.

It was the tenth voyage on the same old wicked track of the sea, with the *Mount Pleasant* three weeks west-bound out of Whydah well slaved and running handsomely, and Downie some five weeks ahead of her, perhaps already cleared-out from Barbados. In the brightness of noon, the look-out had hailed from the mast-head: 'Ship fine on the starboard bow!' but Matthew, peering through his telescope from the lower level of the deck, could see nothing, He sent another topman aloft with the order:

'Take a look, quick as you can. Make out her rig, and try to judge her course. If she seems a warship, hail me back straightway.'

Presently the topman, Evans by name, returned with his report. Being a Welshman, he had a lust to play the theatricals

whenever he could, and this time his voice and bearing took on all the darkling mystery of great drama. He began with the Pause Pregnant.

'Well?' Matthew asked irritably, when he had waited as long as he could. 'What did you make out?'

'Nothing in this mortal world!' Evans declaimed.

'What?' Matthew said. 'What foolery is this? How is she rigged?'

'No rig, sir'—and no voice more laden with doom had ever issued from the grave.

Matthew turned to Harkness at his side. 'Is this man drunk on watch-duty?'

'I'll discover soon enough,' Harkness said grimly.

'Not drunk,' said the great tragedian. 'Call me what you will, call me messenger of evil——'

'Enough, enough!' Matthew shouted. 'I'll call you cook's boy if you play the fool with me . . . Now, listen. What course is she steering?'

'No course, sir.'

'Now hark you here——'

'T'is true,' Evans said, by now well launched. 'I swear by Almighty God, she has no sails, she has no course, *she has no men*! Would you call a blessing, sir, before we sail into the pit? Shall I lead the men in a Song of Praise? Wicked sinners we all be, but——'

Matthew had barely time to think: God's mercy, all we lack now is a Welsh tenor in full flood, when it all came true.

The ship could now be glimpsed, between one wave-top and the next, from the level of the deck, and she was no ship at all: merely a wallowing hulk, a wreck with two jagged stumps for masts, and a tangle of ropes and sails still tethered to one side, already turning to seaweed where they touched the water.

This bleached and blistered corpse would never put to sea again. Her stem sagged deep in the water, as if she were rooting for her grave. Near to, she had the smell of a slaver also to add to a picture of the damned.

She had something else as well. 'By God!' Matthew whispered to himself, as he saw the tattered rags of a house-flag among all the wreckage trailing over the side. 'Can it be true?'

Harkness had seen it also. 'T'is one of our own, sir,' he called from his station on the taffrail. Then half of her name could be read, on the stern canted high in the air. 'Mersey,' he spelled out, with chill horror. 'Yon's the *Blessing*!'

Captain Downie's own.

While the *Mount Pleasant,* her headsails trimmed aback waited some two hundred yards off, Matthew himself boarded the wreck, in a ship's boat manned by six volunteers and the Gunner. He led the boarding-party because it seemed fitting to do so, as the head of a family might attend a funeral to pay his respects for all; and volunteers were best for such a gruesome task as this might be.

He gave only one order on their short journey across, dismayed by the foul stench now pouring from the deserted ship:

'Steer round the bows, and then up to windward, or this will choke us.'

There was no movement on deck as he approached, save for the clouds of flies which had laid a droning canopy over her whole length. But now there was one body to be seen: a skeleton hanging spider-like from the wheel, as if the man had been crucified upon it, and still claimed this post of honour. Matthew scaled the ladder first, followed by the Gunner, though within a moment he could see that there was no gunner's work to be done here. There was nothing to be done, except to make the dreadful climb downwards into an open coffin.

Whatever had struck the *Mersey Blessing* had left a bloodless battlefield, none the less terrible for its silence and its lack of wounds. There were many dead on deck, strewn about in attitudes of peace; but no weapons remained, no bodies hacked nor faces contorted, no reminder of violence. All the dead men were white, and most lay on their backs with their eyes open —staring socket-eyes, cooked to nothing by the sun.

They lay at peace on their ruined deck, a deck carpeted with noisome flies. Every last hatch or entrance-way which broke it was battened down, or boarded over.

The fearful stink proclaimed that the ship still had her slave-cargo below. So the enemy must be storm-damage of an extreme kind; the ship had been caught under full sail, hammered flat upon a boiling sea, and, losing masts and sails, had righted herself . . . Why had such a violent storm not been foretold? The horizon must have been as black and livid as pitch. Had they not seen it?

There was one answer, Matthew thought, which trimmed off all the others. They could not.

The boat's crew talked as they made their discoveries.

'No attack,' the Gunner said. 'No slave revolt or riot. Not more than fourteen men left—not even a full watch. They did not fight each other. Where are they, for God's sake?'

'Swept overboard?' a voice hazarded.

'If the storm was so quick as to take out the masts, then they would still be below, and come up when she righted.' The Gunner said again: 'Where are they?'

'Not a sail remaining that's worth calling a sail,' the bosun's mate pronounced. 'Nor a rope they could use again. The shrouds torn out . . . Look at those dead-eyes—split like oranges! They must have been stripped bare in half a minute.'

'T'is the plague.'

'T'is sea-serpents.'

Thank God that Evans the Prophet was not among them.

'What's to do, sir?' one of the oarsmen asked, and there was silence again.

Matthew was ready with his cure, as any ship's captain must be. 'Nothing but burial,' he said hardly. 'Burial of all . . . She will never sail, and we cannot tow a lump like this half-way across the Atlantic. She is too foul to be touched, in any case. I know the lok of these men. They were blind, and the source is still below . . .' He let this thought possess them for a moment, in case he needed friends in the future. Then: 'I must have the log, and any trading-money on board. Then we'll fire her. Gunner!'

'Sir?'

'Have you powder?'

'Small keg, sir.'

'Lay a slow-match trail from the bulwarks down to the captain's cabin, where I shall be. Plant the keg against the bulkhead. Back it with furniture and any loose timber you can find. Then wait the word from me.' He beckoned to the carpenter.

'Bring your tools and come with me. The cabin is likely to be locked.'

All the horrors of the dead ship now shrunk down to one small cabin and one smaller man. The barred door broken open, they stod back before a blast of foul air and a fresh swarm of flies. Then Matthew, summoning some necessary spirit, crouched and went in.

He saw what he had expected to see: the Captain enthroned in command, but powerless—the dead king of all his other dead. In the fiery heat of the locked cabin, Downie seemed like the ashes of some pitiful sacrifice. He had been seated at his desk; now—and how long was it since his dreadful fall?—his parched bones had pitched forward, spread-eagled like the man at the wheel, secure for ever in his last station.

The struggling grey hairs on the yellow skull lay like some bedraggled nest in the curve of his left arm. Under the other was the log, which Matthew must have. At his back, lashed and chained to the bulkhead, was the money-chest, the other trophy to be taken from him.

Matthew knew a moment of fear and horror, and could only ask himself, Why? He had seen a thousand skeletons. Why should this one choke him? It was no saint, nor even a monstrous sinner. It was only Downie, odious in life, repellent in death. Who could mourn the little monster?

One other man, it seemed, could. The carpenter, a quiet Irishman of middle-age whose customary conversation was of his grandchildren, doffed his cap and said:

'God ha' mercy on him!'

It would do for them both. Matthew said: 'Aye to that,' and then: 'The money chest. Break it loose, but do not open it. Cut the lashings, and strike off the chain staples.'

It was done, with a swiftness almost ferocious, which showed that the Irishman, never one to challenge trouble, would be glad to be gone. Then Matthew told him:

'Send a man down to carry it back to the boat, and bring me a good piece of sailcloth.'

Now he was alone, and now was the turn of the log. Matthew pulled at it, but it would not budge; he was forced to move

Downie's right arm, which slipped across the desk and into his lap with a rattle of dry bones, like dry ice shivering as it broke. But the log was free, and could be lifted, and its cracking parchment leaves closed, and put on one side. Then his helpers returned the Gunner with his powder keg, a strong-armed porter to bear the chest, and the carpenter with his sail-cloth.

None of these hardened men looked at Downie for more than a glance but went to work in silence. When the log was wrapped, they were set free. The charge exploded as they were half-way back to their own clean ship, and within a few moments the fire thickened and began to take hold of the one they had quit so readily.

[Nicholas Monsarrat's definitive text ends here. The remainder of the story has been told in his own words, partly from some working notes that he left and partly from his synopsis of *The Master Mariner* Book 2 which continues the adventures of Matthew Lawe on page 142.]

Log wrapped in sail-cloth, cooked in a galley till the brittle pages could not harbour infection. The flicker (astern) of the funeral pyre as he opened it.

Reads log, dated, and interjects his own commentary.

Slaves had a few cases of ophthalmia, then many, then all. Men would not feed them. Fearful scenes of fighting as blind men grapple and kill. The rotten meat is locked under hatches, too dangerous to throw overboard.

Then the first white casualty. Downie can still see, and write (he has had a dose of this before). But scrawl affected by drink. Finally he is left with seventeen men, blind or half-blind. Others have blundered overboard, or jumped: mad with thirst for water they could not find. Tries to shorten sail on approach of storm. But who can work? One poor wretch, shrieking for water among heartless shipmates, had his head crammed into a shitten-butt and was told 'Drink and be damned!'

At desk, began to rub his eyes (log has become a hysterical notebook). The eye he rubs is immediately clouded over.

*

Makes last entry. Within two days, stone blind to the last man. They were revenged....

At the end of this trip, Lawe has his customary interview and accounting with Benjamin Boothroyd. By now he knows almost every detail of the illegal trading of the past five years, and of the sort of profits which the ship-owner has made; and he cites the disaster to the Mersey Blessing as the sort of hazard which any captain may have to endure.

While the owner sits comfortably at home with his handsome wife and innocent children, he is battening on the courage of a few men and the misery of countless others. Lawe demands a bigger share of the profits, since he takes a far bigger risk in terms of life and limb. Otherwise, he threatens, he will tell his story to the authorities, and sever a foul connection and a fouler trade.

Boothroyd, though furious at this blackmailing demand, conceals his feelings. It is clear that Lawe knows too much about his affairs, and his guilt, and his double-dealing, and is in the mood to use this knowledge. He temporizes, promising another meeting next morning 'when they can sift this little matter out'. Matthew Lawe returns to his lodgings.

It is the last he sees of Boothroyd, or the Mount Pleasant house, or the pretty wide-eyed children whose first question is always: 'Did you have adventures, Captain Lawe?' He is betrayed to the naval press-gang working down in Liverpool Docks and forcibly held as he leaves his own front door. (The war is on again, in 1812, and England once more needs sailors: it is the usual see-saw of full employment and insulting neglect which, with every war, costs so much hardship, so much blood and treasure, before the nation catches up with its peril. This time the urgent aim is to enforce a world-wide blockade against the French, and particularly a trade-embargo against the Americans.)

With his whole freedom and future at stake, Matthew might have bought himself out of trouble, since the press-gang operates a sort of protection racket as well; but when he offers ready cash to his captors, his lodgings are raided and all the hoarded savings of five years' fierce endeavour are stolen.

'We found nothing!' the press-gang chief says, straight-faced. 'What was that foolish talk of a strong box? There was not so much as a mouldy sea-chest!' Lawe is frog-marched away, and within the hour finds himself a seaman on board the Shannon *frigate, bound for Halifax, Nova Scotia.* He had learned nothing.

Envoi

The Continued Story of
Matthew Lawe

nine

Pressed Man
1813

1 **Main Identification** The duel between the frigates *Chesapeake* and *Shannon* during th Anglo-American War of 1812.

2 **Locale** North American waters, Boston.

3 **Principal characters** Captain Philip Broke of the *Shannon* (British), Captain James Lawrence of the *Chesapeake* (American).

Captain Philip Broke was a somewhat eccentric naval firebrand who incurred the wrath of Their Lordships of the Admiralty by holding daily gun-drills on board the *Shannon*, thus exceeding his authorized monthly ration of powder and shot. For this display of zeal he was banished to the equivalent of Siberia—the Canadian port of Halifax, centre of British naval activity in North American waters.

Captain James Lawrence, billed as the tallest man in the U.S. Navy, had a distinguished career in small ships before taking command of the *Chesapeake* frigate. The two men were evenly matched in skill, energy, and foolhardy courage. Philip Broke had the edge in brilliant tactical thinking.

4 Narrative Line

Britain, at a new peak of power, can now call on 1,100 ships and 140,000 men to man them. Approaching the last crunch of the Napoleonic wars against France, she can and does blockade every harbour in Europe, in the effort to starve out her prime enemy.

Presently this conflict escalates into a full-scale war with America, which is trying to break the blockade by the use of 250 'privateers' (licensed pirates based on the earlier model supplied by Sir Francis Drake). America also plans to invade Canada, our last toe-hold on a mutinous continent.

The frigate *Shannon* takes part in this blockade; and Matthew Lawe, as one of her gunners, witnesses a classic sea-duel, which only lasts fifteen minutes, between the two 38-gun frigates, the *Shannon* patrolling the port of Boston and the *Chesapeake* bottled up in harbour. The latter is taunted into leaving shelter, and meets with swift disaster.

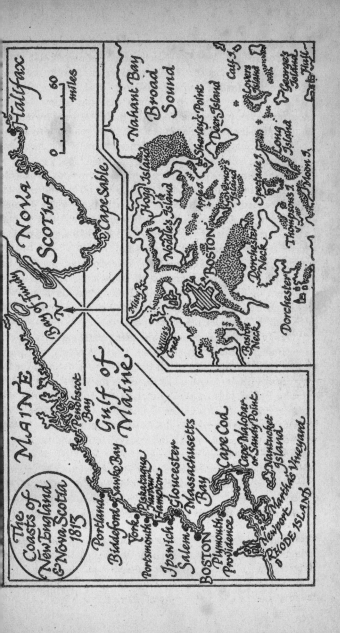

The Coasts of New England & Nova Scotia, 1813

MAINE

Bay of Fundy

Nova Scotia

Halifax

0   60
miles

Penobscot Bay

Gulf of Maine

Portland
Biddeford
Sawko Bay
York
Piscataqua
Portsmouth
Hampton
Ipswich
Gloucester
Salem

Massachusetts Bay

BOSTON
Plymouth
Providence

Cape Cod

Cape Malabar
or Sandy Point

Nantucket Island

Newport
Martha's Vineyard

RHODE ISLAND

Cape Sable

Merrimac R.

Mistick R.

Hog Island
Noddle's Island
Apple I.
Governor's Island

BOSTON

Dorchester Neck

Dorchester Bay

Boston Neck

Nahant Bay

Broad Sound

Shirley's Point
Deer Island

Spectacle I.

Long Island

Thompson's I.

Cats I.

Lovers Island

George's Island

The Castle

The taunting is deliberate and skilful: Captain Broke is determined to find a way of enticing the *Chesapeake* out. After repeatedly showing the British colours off Boston, he sends Lawrence a formal combat challenge in delicately insulting terms: 'Only by repeated triumphs can your little navy console your country for the loss of its trade.' James Lawrence, a huge quick-tempered man, takes the bait; goaded into action he puts to sea, with an inexperienced crew which has never seen action, and runs his neck straight into a bloody noose.

The *Chesapeake* is subjected to two ferocious broadsides (the *Shannon's* illicit gun-drill thus paying off a hundred-fold) then boarded and captured in a fifteen-minute fight. Captain Lawrence, shot through the lung and dying of this wound, gasps out another phrase destined to become a part of U.S. naval history: 'Don't give up the ship!'

But the *Chesapeake*, in this gory encounter, has lost 146 men dead or wounded, against the *Shannon's* 72; and at the moment of Lawrence's death she hauls down her flag. She is escorted into Halifax harbour by her victorious captor, flying the Union Jack over the Stars and Stripes.

It is an immensely popular win, after many earlier setbacks: the 'Ballad of Brave Broke' is No. 1 on the charts, surpassing even 'Rule Britannia', though Philip Broke is disabled for the rest of his life by a cutlass slash across his skull.

The British burned Washington in the following year, and put an end to all U.S. hopes of annexing Canada. After that Britannia rules the waves, almost without dispute, for the next hundred years, until 'your little Navy' took over the role of policing all the seven seas.

The jubilation in England over Broke's victory is summed up by a rude cartoon which anticipates a later phrase. The caption is curt:

John Bull to Brother Jonathon: 'You may kiss my taffrail!'

ten

Foremast Hand
1839

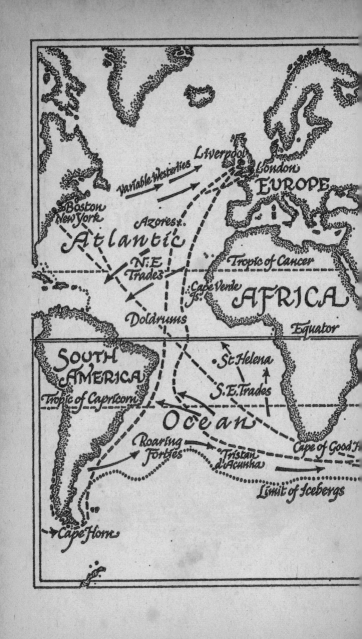

Chart showing
the tracks of the
Clipper Ships

CHINESE Peking   Pacific
EMPIRE
        Shanghai   Ocean
        Foochow
HINDOOSTAN Canton
Monsoon   N E Trades
to April   Hong Kong
Monsoon
May to Sept.

SUMATRA
Indian
Monsoon   Sunda
to Mar.   Straits   JAVA
Mauritius   S.E. Trades
urbon
        Ocean   AUSTRALIA   South New Wales
                        Adelaide   Sydney
                                        New
Roaring Forties   Ulster
                                        New
                                        Munster
Kerguelen

1 **Main Identification** Clipper ships: the great Tea and Grain Races from China and Australia.

2 **Locale** Liverpool, China, Australia, Cape of Good Hope, Cape Horn.

3 **Principal characters** Herman Melville, clipper-ship captains, famous early clippers such as *Flying Cloud* and *Star of Sumatra* (1851 & 1853), a Chinese tea-house girl at Foochow.

4 Narrative Line

Matthew Lawe is 'down again', in the sense that he is once more a common sailor, working as hard as he has ever done in his life, in conditions of fearful hazard aloft and sordid chaos between decks. Though the naval floggings are past history, 'hazing' (bullying by overwork) is still rife in the clipper ships, and a dose of 'belaying-pin hash' (a beating with a thick wooden cleat) is still the medicine for any sort of slackness or insolence.

Two years into the reign of Queen Victoria, it is a moment of driving ambition for Britain, seeking to grab the world's trade routes (as well as the prime cuts of Africa) before someone else gets there first. The East India and the Hudson's Bay companies are both in full flower; and private shipping lines are multiplying daily, giving a solid base to this greatest era of British expansion. It amuses Lawe to recall that the East India Company was originally founded, as the Levant Company, with a minor part of the profits of Drake's round-the-world voyage in 1577.

The first-line weapon of this trading explosion is the clipper ship, patterned on the Baltimore clippers built in America *circa* 1800. The clippers have developed until they can be called 'by far the most elaborate structure and the most complicated mechanism which the mind of man had yet evolved'. They are also supremely beautiful, as they spread their feathers to a snowy mountain of canvas; but they are driven frantically hard, and the men who serve in them are driven hardest of all.

It is the dawn of 'iron men in wooden ship'; and, though

whaling is also reaching its peak, spurred on by a curious twin demand for lamp-oil and corsetry, the men of the moment are the deep-water sailors who man these ocean greyhounds, cramming on sail to make 17 and 20 knots to race their rivals home and capture a bottomless market.

To serve this harsh purpose, they must become used to working in all weathers: the extremes may be a shrieking Cape Horn gale when a sail can be ripped from its bolt-ropes and vanish without trace, or a man crusted and caked with icicles can have his fingernails torn out as he fights, a hundred feet above deck-level, to control a sail which *must* be furled if a topmast is not to be lost: or alternatively, he can be sweating in the tropic heat of the Horse Latitudes, which earn their grisly label from the horses and cattle which have to be thrown overboard to save water.

Here he can be turned out fifty times in a working day, to hoist or trim sails which may catch a chance breeze; or, toughest of all, a man can be sent off in one of the long-boats, to tug and gasp at the oars for hours on end, towing the deadweight of his clipper ship towards a ruffle of water which may presage a change of wind.

Back on board, he is sustained by odious food served in greasy wooden tubs (called mess-kids); the 'choice' is salt pork which can induce a raging thirst, or cracker-hash, a mixture of broken hard-tack biscuits soaked in water and mixed with salt beef and onions. 'Liverpool sailors know this delicacy as lobscouse.) The beef, which is bright red, comes in 300-lb casks soaked in brine and salt-petre, and re-soaked at sea in a 'harness cask' topped up with sea-water. To wash it down, a mug of coffee sweetened with molasses is the only thing on the wine-list.

Since fresh water is more precious than anything on board, a barrel of urine is kept on deck, in which he may wash his clothes. But in cold wild weather his clothes are never dry (nor taken off); and in a hot climate his only home, the sweating foul-smelling fo'c'sle, is uninhabitable.

Such is the human basis of the burgeoning British Empire; and such is Matthew Lawe, veteran of many such crude and killing voyages, as we sight him in the usual waterfront tavern on the usual sleazy Liverpool street. Dressed in his best rig of blue pea-jacket and stained neckerchief, he is 'between ships'; his money is running out, as always, and when he eyes a dockside whore with a bustle as big as a brewer's horse, he knows he cannot afford even this modest refreshment.

Tomorrow he must sign on again, in a good berth if possible, or else the loan-sharks and crimps will beat him insensible and deliver him like a sack of mouldy potatoes to any hell-ship which, by its very reputation, cannot find a crew by any other means. Tonight he sips weak 'four-ale' beer, and counts his pennies. Tonight, probably his last night on dry land, he waits for a shift of wind, and knows that it cannot come. But tonight, for a change, he meets a sailor in the same sort of fix, who is *not* despondent.

He is a tall, black-haired, gangling young man, not much more than a full-grown boy. But he comes bursting into the tavern as if blown there by a fortunate breeze. His eyes dart round the cheap bar-room. He has to talk to someone, and that someone is Matthew Lawe. His name, he says, is Herman Melville.

'Herman?' Lawe is not exactly in the mood for a bouncing young stranger. 'So you're a squarehead?'

The boy grins. 'Hell, no! Yankee! What will you drink?' He obviously has the last of his money to spend, but his purse is a bit longer than Lawe's. 'What would you say to blackstrap? Can we get blackstrap in this benighted place?'

'Depends what it is.'

'Rum and molasses, with a dash of vinegar.' He sees Matthew hesitating. 'Don't you fret, old fellow. It's my corner. Put away that mug of slops. Have a real drink!'

He gets his blackstrap, and shares a pouch of what he calls 'Negro-head—real tobacco! '—pressed and rolled into a hard cake, stiffened with molasses also, needing to be pared with a knife which is instantly ready. Over a strong pipe and a stronger drink, Melville's wonderful life story comes tumbling out.

He's a sailor—well, he has just made his first voyage, New York to Liverpool in a darned old tub of a packet-boat. He is twenty. His father died, bankrupt, when he was thirteen. He has been a bank clerk, then a schoolmaster, now a sailor. It's the sailor's life for him!

But he also wants to write about it, get it all down on paper, show the world what it's like to be a sailor, and make a bit of hard cash on the side. There are stories to be told about the clippers—he doesn't know anything about Yankee clippers but he knows the stories and the men. Men like Captain Bully Waterman, who used to padlock the topgallant hailyards before he turned in for the night, so that no damned coward of a mate could shorten sail. Men like—oh, there's no end to them!

But especially whaling—now *there's* the life! Whaling and writing are the two things Melville is mad about. And blackstrap—let's have some more blackstrap before the shiners (coins) run out. The two of them will have to sign on tomorrow, anyway. What are the chances? Does Matthew Lawe think he can pick up a berth in a whaler? Bound for icy Greenland, or the warm Pacific? How about that?

Matthew is dubious, and discouraging. There's not much whaling out of Liverpool—quick returns is what the owners want, not three-year voyages, with the chance of a cargo of empty barrels at the end of it. Beter to go back to a port like New Bedford in Massachusetts, where there's a huge fleet of Antarctic whalers. But why does Herman Melville want to go whaling, anyway? It's the worst life in the world, and the most dangerous. Hasn't Melville ever heard of what happened to the *Essex*?

Melville hasn't heard (he was one year old at the time), and for a change he listens as Matthew Lawe spins a cautionary yarn. The whaler *Essex*, 238 tons, was rammed and sunk by a huge rogue whale in mid-Pacific, and the survivors (8 out of 20 men) spent 96 days in open boats before being rescued. In their wanderings, they avoided certain Pacific islands because of their fear of 'cannibal savages', but in the end they all ate their dead comrades in order to stay alive.

One of these survivors, Captain Pollard, was interviewed by an innocent reporter sometime afterwards. There had been some talk about the cabin boy of the *Essex*, the youngest to die. (He drew the short straw, and was killed for the pot. His name was Coffin.) He must have been a promising lad. Did Captain Pollard know him well?

'None beter!' said Pollard. 'Hell, son, I et him!'

Thus (though Matthew Lawe only realizes it later) is the genesis of 'Moby Dick' planted. That night, he and Melville see the tavern closed, and agree to meet on the morrow, to sign on for the best ships they can find. Melville never turns up, and Matthew Lawe goes through the traditional ritual of choosing and joining a new ship.

For this, any dockside tavern will do; it is the luck of the game. Lawe picks on the tap-room of the 'Ferryman's Arms', and finds it crowded with sailors come to see which ships need crews and what berths are open. The men are of every sort: they range from skilled seamen—and proud of it—who can 'hand, reef, and steer' with the best, or carpenters and riggers and sail-makers who may have to repair a mast-less, stricken ship in mid-ocean, down to the shirkers and wharf-rats who drift from ship to ship like the weed fouling its bottom, the sea-lawyers who whine interminably, the cooks who, in the crew's derisive phrase, cannot boil water without burning it.

They all carry their discharge papers from their last voyage; and they wait patiently as the trestle tables are set up in the tap-room. Behind each sits the mate or the ship's agent who is to engage a crew. It is an important moment: on it will depend a man's happiness or misery for the next year or more, or even his chances of life and death.

The names and berths are called out: 'Six A.B.'s for *Bonaventure*!' 'Cook and carpenter for *Barra Head*!' There is some murmuring, and a mortifying lack of volunteers, when certain ships are named. Bad ships are poison, and such news travels fast along the waterfront. These are the ships which, in the last resort, will have to take the drugged and drunken sweepings of the doss-houses, delivered on board by the crimps and pimps and sharks of the dock area, who swindle a rake-off of the first month's wages, by papers often forged, or 'signed' with the thumb-mark of a half-insensible man.

But in the more orderly atmosphere of the 'Ferryman's Arms', there is a slightly better taste to employment. When a crew has been assembled, they move to another room where the 'Articles' which set out the terms of their service, and the disciplinary powers of the captain, are read out to them. The last three words 'NO SPIRITS ALLOWED', are always enunciated with special clarity; the rest is taken at a fast gabble.

Then they sign, or make their mark, and receive a note for a month's pay in advance; the proceeds of this note, cashed at a 40% discount by dockside money-lenders, is usually drunk there and then. It is redeemable three days after the ship clears South Stack lighthouse (Holyhead, 'provided the said seaman sails in the said ship.' The shore-side bully-boys make sure that he does.

Matthew Lawe goes through this routine, and thinks he is in luck: A. B. on board the crack clipper *Star of Sumatra*, bound for Foochow to load with chests of tea, and then to race home with her cargo. The enormous snobbery of a new society fad, to drink 'the first spring tea from the first ship home' (at 18/- a pound for prime Pekoe), fuels this contest.

Next morning, reporting early, he watches his shipmates come on board. They arrive in various stage of drunkenness and disorder, hilarity and dejection, song and silence; their escort may be anyone from a boarding-house keeper who swears blue murder that he has not been paid, to a girl trying for a final dividend between the dock gates and the ship's side. Many of the last arrivals are frogmarched or carried bodily over the gang-plank: the tide cannot wait, nor the men who have an investment in their safe delivery.

Most of the crew have at least a battered sea-chest or a canvas kit-bag containing all their possessions; but a few unfortunates arrive in 'parish rig'—the clothes they stand up in. Later they will pay for this privilege, since they will have to buy foul-weather clothing from the captain's slop-chest, at prices suitable to a floating monopoly. At the end of the voyage, with all its sweat and terror, such a man, his pay mercilessly docked, may not have a cent to show for it.

Now, down on the Liverpool waterfront, the gang-plank is

hauled in, the sick and sorry men are dowsed with buckets of raw Mersey water and kicked up the mast to unloose the sails; and the proud (or soon to be proud) *Star of Sumatra* sets out on her 14,000-mile journey to Foochow.

On this voyage, Matthew Lawe knows where he is going, and loves it, though the three months' battle eastwards towards the Dragon Sea is as tough as ever, and the ship is driven, like her crew, to the very limits of strain and endurance.

Their course is a 'down-easter', with the ship bashing her way out into the Atlantic, catching the South-East Trade Winds to 35° south, and then a reliable westerly round the Cape of Good Hope; sailing with an easy wind across the Indian Ocean, and then forcing a passage through the Sunda Straits (between Java and Sumatra) into the typhoon-ridden China Sea. But once there, a short and heavenly calm descends.

Lawe has a girl in Foochow, the 'tea-outlet' of all the Orient: she is a Chinese dancing girl, slim as a pin (in contrast with that Liverpool girl with a bum like a brewer's horse), talented at music and conversation, formal and cultured, giving her favour with rare discrimination—such girls are not active whores, but entertainers who may, out of grace and favour, relax the rules at their own moment of choice.

Some of her serious talk concerns the scandal of the new opium trade, the illegal traffic in this 'foreign mud' which is shaming and degrading China.*

Meanwhile the *Star* is loading with countless chests of the first spring tea: chests so beautifully made that they are often

---

* The tea we bought from the Chinese was expensive, and they insisted on being paid in silver ingots, which were hard to come by. To raise the silver to cure this, our first adverse balance of trade, we had to find another commodity which we could sell to the Chinese. Opium was the answer. It was illegal in China, but this did not stop the East India Company from growing it on their plantations in India, getting it smuggled in, selling it for Chinese silver, and thus paying for the tea.

We had to fight an Opium War (1839–1842) to make sure that the stuff got through without being confiscated and its carriers executed. In fact, it was we who corrupted the Chinese with this insidious drug; and the idea of a 'hideous opium den' as the invention of villainous orientals is nonsense.

used for furniture when they reach England, and lined with paper-thin beaten lead of such a quality that it is sold to newspapers for the type-setters to use. The incomparable Chinese stevedores can load 8,000 chests in fifteen hours. It is the haggling about the price which takes up the time.

On the return journey, the homeward swing of a gigantic pendulum, the captain of the *Star of Sumatra* forces every ounce from his ship, and overplays his hand. The laden clipper, running into 'Bank weather'—a fog as thick as the inside of a glove—in the Irish Sea, and then a gale heralded by a dreadful 'line squall', is wrecked off Ravenspoint (Holy Island) in spite of desperate efforts to claw her off the coast.

Lawe and most of the crew are rescued by the lifeboat from Trearddur Bay, Anglesey, one of the first breed of 'submersible self-righting boats' invented by William Wouldhave, clerk of the parish church of South Shields.

Matthew does a second clipper voyage during this period, 'running the easting down' past the Cape of Good Hope, across to Australia, loading there with grain at a port which has promoted itself from the convicts' miserable Botany Bay to rumbustious Sydney Harbour, and carrying on eastwards round the Horn, across the Atlantic again, and then to 'Falmouth for orders'.

It is a giant 30,000-mile circle of the globe which ends in the East India Dock below the Pool of London—but by now as normal and matter-of-fact as a city gent's visit to the Baltic Exchange in St Mary Axe, where the hard-won grain is sold.

Once more he, like a thousand other sailors in the same half-year, has encompassed the world. But the London papers do not record such run-of-the-mill intelligence. They are full of a *real* story—the smashing of the Portland Vase, old Sir William Hamilton's dearest treasure, by a lunatic at the British Museum.

eleven

Look-Out
1857

1 **Main Identification** The voyage of the *Fox*, to find out what had happened to Sir John Franklin's last expedition (1845) to discover the North-West Passage.

2 **Locale** The Artic.

3 **Principal characters** Explorer Sir John Franklin (in retrospect), Lady Franklin (his wife and widow), Captain Francis Leopold M'Clintock, commander of the *Fox*.

Franklin, a gifted navigator, had one of those remarkable pedigrees which recur so often in the ocean world (and thus in this book). He took part in the Battle of Copenhagen at the age of fifteen, and was a signal midshipman in the *Bellerophon* at Trafalgar. He had made two previous tries (1819 and 1825) at the elusive North-West Passage which still intrigued and fascinated all sailors, before giving up hs job as Governor of Tasmania and setting out again, in his sixtieth year, in command of the *Erebus* and *Terror* (1845).

Above all, he had learned his navigation and marine survey from his cousin Matthew Flinders, famous explorer of Australian waters, who was a pupil of Captain Bligh, who was Captain James Cook's sailing-master.

Lady Franklin, his wife, was a woman of dauntless devotion, energy, and some fortune, who squandered all these in trying to establish what had happened to her husband, disappeared from the face of the sea along with his two ships and his crew. She encouraged, promoted, and in some cases paid for fifteen expeditions to learn the truth: it was the prime motive for Arctic exploration for the next twenty years.

Much of the mystery was cleared up in 1857 with the voyage of the *Fox*, a 177-ton yacht (with auxiliary steam power) which she bought and fitted out at her own expense (the Admiralty being by now somewhat sick of the whole subject), under the command of Captain M'Clintock.

This was the man who picked up the pieces. A determined and dedicated Arctic voyager himself, he proved beyond any reasonable doubt that Franklin had in fact found the Passage, and had perished after incredible hardships, with no single survivor to tell the tale.

4 Narrative Line

At sixty-five Jane Franklin is a very determined woman, and (one may deduce) still a considerable charmer as well. She had said goodbye to her husband in 1845; eleven years and some fifteen expeditions later, she is still plaguing the British Board of Admiralty to find out what has really happened to the *Erebus* and *Terror*.

She is not satisfied with certain 'relics' brought back so far: she is not satisfied with the Admiralty's contention that the mystery has been 'sufficiently solved', and that it would be wrong to put more men at risk when there can no longer be any hope of saving life: she is not satisfied with the rejection of a 3000-word 'memorial' on the subject, addressed to Prime Minister Lord Palmerston.

Though now reconciled to the fact of widowhood, she is still determined that the Arctic must be made to yield, if not the body of her beloved husband, then at least the place and manner of his death, and, God willing, the proof of his triumph. In 1856 she still has half her fortune left; and with this she buys a large 'screw-yacht', the *Fox*, for £2,000, fits it out (after removing its 'velvet hangings'), assembles a crew of Arctic veterans who must all have been her devoted admirers, and sends them off on yet another search.

It says much for her reputation that applications pour in from all over the world to join the *Fox* 'in any capacity': that certain of the officers chosen, whether impecunious or not, contribute to the costs of the expedition: and that none would take any pay 'if it comes from Lady Franklin's resources'.

She is fortunate in her choice of commander, Captain Francis Leopold M'Clintock. A 38-year-old Irishman of great strength and spirit, he has already taken part in three Franklin searches, including the very first one (1848).

With an eye to the future, he has qualified as a steam-engineer and he has also given great thought and study to polar exploration, with special attention to sledge journeys—the design of lighter sledges, the loads and rations which can be carried, the technique of 'planting ahead' periodic caches of food which can make a long journey into the unknown safer, and a retreat possible. (The American admiral Robert Peary was to reach the North Pole, and Captain Robert Scott the South, by copying these methods.)

Above all, M'Clintock knows, almost by heart, the clues he has to work on. They all point to his favourite theory—that Franklin would not seek a passage in the impenetrable 'far north', but would follow as closely as possible the North American coastline, the last 'land' before the wilderness of ice takes over. Here, if anywhere, is the only navigable water.

Over the years, the evidence has mounted up. At first it is 'a charred board' in one channel, 'a piece of wreckage' in another. Then explorer Erasmus Ommanney (1850), with M'Clintock among his crew, finds the remains of a camp-site on Beechey Island (to the north, at the natural entrance from Baffin Bay): there are rags, bits of rope, and three graves.

Franklin has wintered here, and there is evidence that he left in a great hurry, with no message to say where he was going— an unheard-of oversight among such men. Then, in 1854, proof of disaster has to be added to the facts.

This comes from another noted explorer, John Rae, who has voyaged down south (where M'Clintock now knows the trail must end), and found on King William's Island, right on the American coast-line, some Eskimos who had seen a party of forty white men dragging their boat across the ice.

When was this? Four years ago, six years . . . Where are they now? They died here and there, somewhere to the south. As proof of this, Rae is offered something the Eskimos would like to sell, or exchange for pemmican or needles. It is Sir John Franklin's Order of Merit.

So the dead have been robbed. But where exactly are the dead?

Now, at last, it is the turn of the *Fox*, and M'Clintock, and Lieutenant Hobson the second-in command, and quartermaster Matthew Lawe. Like M'Clintock, he has been here before—with Henry Hudson in 1610. M'Clintock, from past

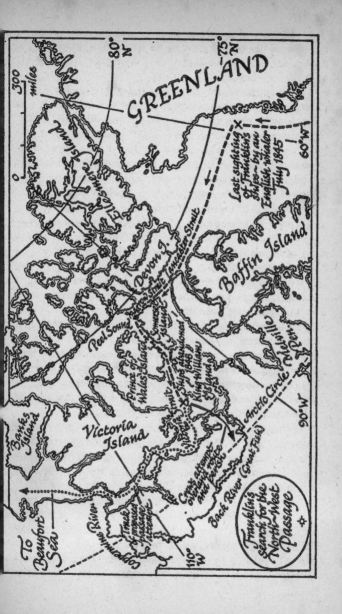

GREENLAND

80° N

75° N

300 miles

0

Ellesmere Island

Last sighting of Franklin's ships—by an English whaler July 1845

60° W

Baffin Island

Devon I.

Peel Sound

Lancaster Strait

Somerset Island

Prince of Wales Island

Prince Regent's Inlet

Victoria Island

King William Strait 1848

Melville Pen.

Arctic Circle

90° W

Banks Island

Crew's attempt overland to Back's route

Back River (Great Fish)

Coppermine River

To Beaufort Sea

Franklin's last composed passage attempt

110° W

Franklin's Search for the North-West Passage

evidence, has already made up his mind where Franklin is *not*
Now all the evidence points to where he must be: on King
William's Island, or *across the North-West Passage* on the
American mainland.

So it is proved, though slowly and painfully. The *Fox* has a
long and tough journey even to approach her Arctic target:
no tougher than for other like-minded ships and men, but
punishing none the less. She is 'beset'—locked in—by the ice of
Greenland, and has to spend her first winter there: a costly
waste of eight months' time, passed in sledge-drill, dog-training
building snow huts, bear and seal hunting, and billiards
played on a table made out of a block of ice ('the bubble
pricked'), with walrus-hide cushions.

There is also plenty of 'Allsopp's stoutest ale', which
M'Clintock, an inquiring man, notes will freeze almost instantly at minus 35 degrees, but thaws again at plus 22, when it is
'muddy but drinkable'.

The first spring releases them at last, after a nerve-racking
12-hour escape journey through monstrous, heaving pack-ice
over tumultuous overfalls. But anything is better than to be
beset a second time. One ship they knew, which did not seize
its first chance, moved a total of seven miles in two years.

Of course, their ordeal has scarcely begun. Clear of the ice
they push on westwards through furious gales, past Baffin
Island to Beechey, and then turn towards their goal, King
William's Island, nearly 500 miles to the south. But in spite of
superhuman effort they cannot get there.

Shifting ice and wicked weather bar the way. Their ship runs
aground, and is thrown on her side, while all round them
giant icebergs are 'calving'—breaking up into huge splinters,
each the size of a house, which hurtle through the air and
crash down like a thousand glittering daggers.

They run short of provisions, and the Arctic game seems to
have deserted this wasteland. There is an average weight-loss
of 35lb per man—about 2½ stone. They lose three men, one of
them through scurvy, which renders his body so brittle that
when he falls on the ice, his bones shatter like glass. Another
dies of apoplexy and a third, the 'engine-driver' (chief stoker)

falls down an open hatch, and is buried in an ice-tomb under a spectral moon.

Then the *Fox*, driven back countless times by the autumn gales, has to give up, and settles down for her second icy winter (in Bellet Strait), still 200 miles from their goal. The real journey has still scarcely started.

As soon as they can, in the weak and watery sunshine of February, they begin to blast their way out of captivity, with axe and saw and charges of gunpowder. But it is clear that a ship—this ship—is not strong enough to force her way through. In April of 1859, leaving only a ship-keeping party on board, they take to their sledges, and split up into four parties, and begin casting about. Already the *Fox* has suffered a great deal, but not as much as Franklin's *Erebus* and *Terror*—for soon a party of Eskimos gives them the news they most fear.

The two ships they seek are no more. One was 'nipped by the ice', and sunk, the other driven ashore and presently ransacked. Both ships were empty. The Eskimos tell their story unwillingly, perhaps with good reason. For the ships are gone, the men are gone; and only the Eskimos have any profit to show for it—more 'relics', more souvenirs, and no single strand of hope or humanity to balance the account.

From now on it becomes a pitiful detective story. Piece by piece, the men of the *Fox*'s sledge-teams assemble the dreadful truth.

By far the most definite, the most detailed, and the saddest news is found by the search-party of which Matthew Lawe is a member: a team of six men altogether, with two sledges, led by Lieutenant Hobson, M'Clintock's second-in-command. Their set task is to search the whole southern and western side of King William's—a hundred miles of fearsome, craggy, howling wilderness, the very coast where *Erebus* and *Terror* were lost. As they advance, they gather like scavengers the appalling fragments of disaster.

There are scraps of clothing, discarded equipment, splintered sledges, and face-down skeletons, all the way. The evidence is of a ragged retreat southwards, which ended like a parched stream running out. Only the foxes—and perhaps some Eskimos—have attended this wake.

163

After many days, Matthew Lawe sights a stone cairn on a hillside. Together he and Hobson scrape and dig and crow-bar their way to its heart. There they find the only written evidence of what has happened to *Erebus* and *Terror*, and Sir John Franklin's only epitaph.

It is a lithographed Admiralty form, and there can be no more pitiful example of how a dry-as-dust clerk's brain-child can be transfigured. In six languages, it enjoins the finder to forward 'this paper' to the Secretary of the Admiralty, London, or to the nearest British Consul, *noting the time and place where it was found*. But round and round the margins of the formal print, men writing in God-knows-what torment have told what they can of their own fate.

In September 1846 Franklin, making for the coast of America, had been pressed ashore by a torrent of ice coming down from the north. He had died in June of 1847, joining the twenty-four men of his expedition already lost. The two ships were deserted ten months later, after being beset for nearly twenty months. Captain Crozier, commander of the *Terror*, and 105 souls, had landed five leagues north of the cairn, and had 'started on'—in April 1848—for the Great Fish River on the American coast.

M'Clintock, brought up to date by Lieutenant Hobson (who conceals the fact that he has himself contracted scurvy), turns all his efforts southwards, to follow a trail already eleven years old. It is still a detective story, but now the heart has gone out of it. All they are doing is taking the inventory of a vast 'Estate of Sir J. Franklin, deceased'.

The Eskimos, who have been inclined to secrecy and a certain evasion, like any other witnesses who 'don't want to get involved', now begin to talk more freely. They make it clear that one by one, after their third winter, Crozier's men must all have perished of starvation or scurvy. A woman tells of sailors 'who fell down and died as they walked along the ice'. An old Eskimo leads the searchers to a bitter winter landscape: a crumbling boat on runners, a gun leaning against it, two skeletons attendant.

Some of the last Crozier survivors had been reduced to eating *tripe de roche*, which sounds better than lichen. There are chewed scraps of roasted leather from mouldy dog-harness, and the burnt bones of prey abandoned by the wolves. Last of all is an Eskimo story, of a party of forty men who had lain down and died, in a cave at the mouth of the Great Fish River —on American soil, across the channel leading westwards. (It is now identified, on very large-scale survey maps, as Starvation Cove.)

The account is closing. It seems probable that Captain Crozier and the last of his 105 souls, limping and stumbling southwards, had hoped to find a Hudson's Bay Company trading-post. They did not. The very last handful found the North-West Passage, and all found death.

It is proof enough that Franklin's men reached their goal— and so, to them the honour! It can never be taken away from them, nor they from it.

M'Clintock, painstaking to the last, and a mourning executor as well, makes a minutely-catalogued list of the 'relics' collected (and bought) from the Eskimos, or found by his own exertions. Apart from skeletons, there are thousands of items, from 'Four cakes of Navy chocolate' to 'The bung-stave of a marine's waterkeg': from '2 small pork rib-bones' to 'One farthing'.

None is sadder than any other; but perhaps 'A table fork bearing the Franklin crest', 'A pemmican tin, painted lead-colour, marked "E" (*Erebus*) in black', and 'Fragments of a boat's ensign' are, for Captain Francis M'Clintock, the proud heart of the matter.

twelve

Supercargo
1893

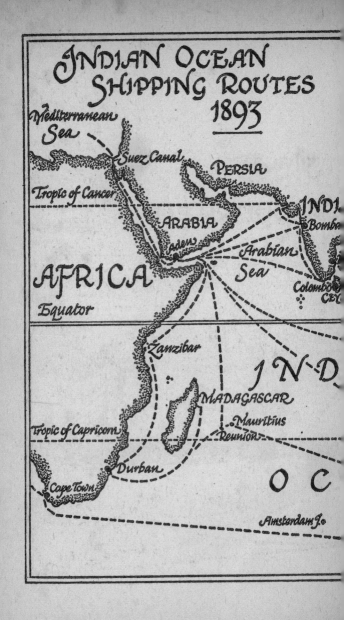

Main Identification The peak of sea-borne commerce under Pax Britannica, as sail gives way to steam.

Locale Adelaide, Australia, and Bombay, India.

Principal characters Novelists Joseph Conrad and John Galsworthy, Samuel Plimsoll the 'Sailor's Friend', the Captain and passengers of a 'steamer' traversing the new Suez Canal to India, including Alfred Bindle (export/import clerk in Bombay), and Joshua Slocum, first round-the-world solo sailor.

Joseph Conrad, not yet a published writer, is on his last voyage, as first mate of the *Torrens*, a crack clipper ship on the Australian run. None of the crew (or the passengers) really know what to make of him: he is extremely good at his job, an imaginative story-teller when off-watch, but always 'different' —a Polish expatriate who has taught himself English, with a dark-eyed personal magnetism which affects everyone who comes in contact with him.

One of the *Torrens*' passenger is young John Galsworthy, at 25 a barrister who really wants to travel and write. He has just made a romantic pilgimage to Samoa to try to meet Robert Louis Stevenson, but failed—he ran out of cash, and Stevenson is dying anyway.

Samuel Plimsoll was a Liberal M.P. who devoted the whole of his working life to the betterment of the sailors' lot, crusading against their conditions of employment, the odious cattle-ships in which (as in the old slavers) half the cargo of export animals might die on the voyage, and the notorious 'floating coffins'— ancient, unseaworthy vessels, grossly over-insured, which were sent off on what the owners hoped would be their last voyage.

Plimsoll gave us the 'Plimsoll Line', the safe-loading mark on every ship at sea, under the Merchant Shipping Act of 1876, and also initiated a tremendous advance in service conditions, crew accommodation, and proficiency examinations for every officer, from apprentice to master mariner.

Alfred Bindle, one of the 'steamer' passengers returning to India from home leave, is a classic example of the man who is nothing in the City of London, but a demi-god as soon as he is re-installed in his colonial job. He seems to grow perceptibly bigger with every mile they draw nearer to Bombay.

Joshua Slocum did a solo circumnavigation of the world, 1895 to 1898, in the 37-foot sloop *Spray*. *Not* sponsored, he paid his own way by giving public lectures en route, and Matthew Lawe, avidly following his progress, either attends one of them or meets him ashore. Slocum had a memorable encounter with President Kruger, the South African Boer War leader. Slocum said he was sailing round the world. 'How can you claim such a thing?' Kruger asked scornfully. 'The earth is flat!' Slocum's own eccentricity concerned his only navigational timepiece, a cheap American alarm clock, one of the first Ingersolls. Whenever it stopped, he boiled it.

4 Narrative Line

There are two voyages in this section: (a) *Matthew Lawe's last* voyage in sail in the *Torrens* clipper from Adelaide, and (b) his first service in steam, as a supercargo (purser) in a cargo/passenger vessel going to India.

The *Torrens* leaves Adelaide, homeward bound, in March 1893. 'The wonderful *Torrens*', as she is nicknamed, is a great favourite on this run; at 1300 tons, she is one of the best and fastest sailing ships ever built, and at this time holds the record for consistently quick voyages—fifteen trips between Plymouth and Adelaide at an average of seventy-four days each. In Adelaide, she says goodbye to another famous flyer, the *Cutty Sark*, still loading wool for her homeward run.

Matthew Lawe has sailed with first mate Joseph Conrad before; the latter has a great reputation both as a seaman and as a man who is prepared to treat every member of the crew as a separate human being, and, astonishingly, is 'kind and forbearing' to the apprentices. They spend most of the first night fighting a fire in one of the holds—something of which the passengers know nothing till long afterwards.

But it is a slow voyage—four months, under a new and prudent captain—from Adelaide to the Cape of Good Hope, the Cape to St Helena to London; and there is plenty of time for Conrad to entertain the passengers on the after-deck, with the twenty years of land and sea tales which he has stored up.

Galsworthy and the others listen enthralled as this strange little bearded man with the compelling voice and intense manner spins his yarns: of his young days as a Polish revolutionary and gun-runner, of men who crack under pressure, or withstand it, of shipwrecks in mysterious eastern seas, and of the great clipper ships—*Thermopylae, Flying Cloud* which could log 21 knots, *James Baines, Sir Lancelot,* and the *Ariel* and *Taiping* which fought a duel half across the world on the China run, and after four months and 14,000 miles entered the Thames River within minutes of each other.

'But now we are too slow,' Conrad tells them. 'Steam is here and the Suez Canal is here, which cuts the voyage to the Far East by thousand of miles. We cannot use it, because the wind there blows ten different ways, and in the Red Sea it doesn't blow at all. So what can we do to keep up with the steamers? What can we do against a ship like the *Great Eastern*?— 9,000 tons, six masts, four funnels, paddles *and* propellers? *There's* the future! But the past is wonderful too. I remember story——'

There is another enthralled listener, besides the privileged passengers: the man at the wheel, ten fet away, and the man is sometimes Matthew Lawe. He pricks up his ears as he hears Conrad mention Herman Melville.

'A wonderful writer,' Conrad tells Galsworthy, 'though his est book came to nothing. He died a couple of years ago— that's why I was remembering him. He loved and feared the sea, as a man loves a woman whose kiss can only mean sweet death . . . Do you know his story, called *The Whale*?'

Galsworthy doesn't know *The Whale*.

'It's a tremendous story about a great white whale—Melville called him Moby Dick—which all the whale hunters chased for years, and which even turned on a whaling ship and sank her. But there's more to the story than that, much more. think what Melville was trying to tell us——'

Later, Lawe summons up his courage—since Conrad has shown his kindness before—and asks if he can read the story of the whale. He explains that 'he has heard of Melville', but did not know that he had gone whaling.

'You're a strange fellow, Lawe,' Conrad says, with one of his sharpest looks. 'Sometimes I think you've heard of everybody. I wish *I* had . . . Well, you can have the book if you like. But——'

The 'But' is right. Matthew cannot make head or tail of the story, and gives up half-way through. But it is nice to know that Herman Melville got his whale-ship at last.

One thing Joseph Conrad does *not* tell John Galsworthy is that he has a book of his own, half finished, locked up in his cabin desk. Perhaps it is just as well: one can imagine Galsworthy thinking 'Oh God!', and declining the privilege of reading the manuscript. Everyone thinks he can write a book. As a matter of fact, he has a jolly good idea for a book himself . . . Conrad's first effort, *Almayer's Folly*, is published the following year, with instant success, and he leaves the sea for ever. He is now thirty-eight, and, as Josef Korzeniowski, did not speak a word of English until he was twenty-one.

In London, Lawe remembers what Joseph Conrad has said about the future of steam and the dying days of sail. He is not too old to make a change . . . By a stroke of luck, he is able to get in touch with Conrad again. The latter is racing to finish, with great effort, a novel which has already taken him four years of meticulous toil; but he is, typically, generous with help and advice.

Yes, it is a good moment to 'go for steam'. Yes, there are berths to be had, but a man must study and learn in order to deserve them. Yes, he might be able to get Matthew a job as supercargo (or assistant, anyway), if Lawe wants to rise in the world. But *no*, whatever the temptation of high wages, he must not take a job on a cattle boat.

It is here that Samuel Plimsoll comes into Lawe's life, either by lecture or book or chance meeting. As part of his crusade to introduce greedy ship-owners to the word

umanity', Plimsoll has recently published a horrific book, of
:kening detail, about what happens to cattle when they are
ansported by sea. It is normal procedure for them to spend
whole trans-Atlantic voyage, of two or three weeks, standing
• in pens on the upper deck, wedged head to tail like sar-
nes with legs and horns; the lateral space-allowance for each
aimal is 2ft 3in, and when the ship rolls or pitches they can
aly gore each other to death.

Since the importers will only pay for live cattle, the beasts
'e subjected to the most hideous tortures in order to give
.em some semblance of life on the dockside. But there are
nitations. Jabbing with pitch-forks may kill as many as it
ares of lethargy, particularly if an animal is already trailing
; guts in a swill of dung and bloody urine. Tail-twisting does
ot always work: tails have a habit of ripping out at the stump.
ne effective trick is to stuff their ears with straw, soak it in
araffin, and set it alight.

The stock-losses after a stormy voyage can be notable. Plim-
·ll produces some figures: from one ship, 33 are landed alive
it of 680 head, from another, 16 out of 276, from a third, 14
it of 360. The ships themselves, laden with these monstrous
:ck cargoes, become unmanageable in bad weather, and fre-
iently founder. When this happens, some owners tend to
ear down hard on the widows and orphans. They claim that
e dead men were paid their wages in full before the voyage
:gan, overtime included.

And the receipts for this unusual bounty? Alas, they went
)wn with the ship.

Lawe agrees: *no* cattle ships. With Conrad's help, he secures
s berth as assistant supercargo on a smart new steamer, the
'encathra, going out to Bombay.

he dock-wallopers have loaded the last of a multifarious
.rgo. The crew are all on board in good time for the tide,
id if there are any disreputable drunks among them, the pas-
ngers do not see them—this is late-Victorian England, where
ood appearances are everything, and in any case bad be-
aviour among the lower classes is the quickest way to the

175

work-house. The passengers themselves have said their las
tearful farewells, and made their way on board.

They are as varied as the cargo: soldiers rejoining their regi
ments after a spell of home leave, missionaries with a Bible i
one hand and a supply of modesty vests in the other, failure
going out to try their luck in a new country, girls on the shel
at twenty-five looking for husbands in a market-place mor
hungry and more promising than England's: clerks, salesmer
Indian Civil Service officers, 'trade-wallahs' who are th
second-class citizens of a vast Victorian snobbery, professiona
men who are a cut above such vulgar thrusters, titled peopl
who have a divine right to their precedence.

They are all going out to India, the brightest jewel in th
crown of Empire, where, being white, every last one of ther
is superior to the 200 million natives whom they are going t
rule, observe, educate, or rob.

The *Blencathra* is a 'pure steamer'; the days of half-steam
half-sail (when there was a cumbersome naval manoeuvr
known as 'Down funnel! Up sail!') are finished with. She i
propeller-driven, which is also very up-to-date: the final battl
of screws versus paddle-wheels has been settled by a tug-o
war match between (appropriately) two tugs: the paddle
wheel model was ignominiously pulled backwards up the rive
at Sheerness.

Now the *Blencathra* leaves the Pool of London, passing som
of the greatest land-marks of modern sea-faring England
West India Dock, East India Dock, Blackwall Dock, Execu
tion Dock (where pirates were hanged in chains), the Isle c
Dogs, Greenwich, Tilbury, Gravesend.

She also passes the mournful 'hulks'—old dismasted men-o
war anchored in the river off Woolwich, and used to house con
victs awaiting transportation-for-life. They are known t
facetious non-sufferers as 'The Floating Academy'.

The long voyage to India is smoothly made, as if Victoria'
Pax Britannica can control the waves themselves. *Blencathr*
takes on coal, a dirty and exhausting job, at Gibraltar, Malt
Port Said, and Aden. The passengers are in proud raptures a
they traverse the Suez Canal: this modern water-way, whic

lops 6000 miles off the Cape route, may have been built by a Frenchman, but it is British owned, operated, and protected.

The passengers are in proud raptures about almost everything; it is the product of a colossal self-assurance derived from Britain's imperial role, which (to be fair) is based on certain ancient virtues—financial probity, value for money, the-word-is-the-bond, and unremitting hard work. Its link, its life-blood, and its guarantee are all the work of sailors.

Blencathra herself can stand for the ship of state, majestic, formidable, and not to be resisted on land or sea. Britain 'has civilised three continents and educated half the world' (*Churchill*, later): has sent her sons to open up the sea-lanes, from China to Peru, New Zealand to Newfoundland, and to keep them secure under an iron-clad umbrella; and has used despised 'trade' as the sharpest weapon of penetration on the face of the globe. The torrent of humdrum goods for mysterious countries is the formula of unrivalled success.

Matthew Lawe, whose centuries-old story this is, finds first-hand written evidence of this penetration as (his superior falling ill and presently dying) he toils with the cargo manifests and makes arrangements for their 'management and sale' on arrival. He has his own interest in this: every member of the crew above a certain rank is allotted his share of cargo space, and is allowed to trade privately.

The captain's portion might be as much as 100 tons, and the assistant supercargo's two. Lawe, advised by Joseph Conrad, has invested in London-made saddles, shoes and boots, which have a ready market in Bombay and Calcutta.

Even the passengers are allowed into this act: by custom, they furnish their own cabins, with everything from harps to grand pianos, chandeliers to *chaises longues*, and dispose of them on arrival. (They have less freedom with their laundry: nothing, the ladies are warned, is 'so indelicate, indeed so indecent' as to hang their underclothes out of a porthole to dry.)

Lawe spends a lot of time during the voyage with Alfred Bindle, the export/import manager whose firm owns much of the *Blencathra's* cargo. He learns a great deal of trade tactics, of how a rival's consignment may be held up by a little 'palm-greasing' among the stevedores, of who can be bribed among the customs men and shipping clerks, of such refinements as the 'salt-water invoice', concocted at sea, wherein certain wholesale costs can swell mysteriously during their passage.

He is amused by the way that Bindle, the outcast 'trade-wallah', who sits at the bottom end of the third officer's table, becomes Bindle Sahib, 'our esteemed manager' and a local lord of creation, in a spotless white suit and best-quality solar topee enthroned behind a colossal roll-top desk, cooled by a two-man punkah, sustained by 'burra pegs' of whisky, and keeping suitors kicking their heels for half a day in the outer office, as soon as he is ashore.

The power of Bindle, and the power of Britain, is well illustrated by a full-page newspaper advertisement, and a great pile of hand-bills, which Bindle shows Lawe in his office about a week later. It is an imposing list of the treasures which the *Blencathra* has brought out, and which are now 'exposed for sale in all the principal emporia, or subject to negotiation by private treaty'.

The life-blood of Britain flows through strange veins.

A large consignment of liquors – claret, red port, 150 pipes of Madeira, old hock, porter, cider, ale, pale beer, rum and gin.

Pineapple cheese, ham, tongue, confectionery, pickles: pickled oysters, Double Gloster cheeses, spiced beef without bone, carraway wafers, wine biscuits, prime St Lucia coffee.

Millinery which includes dresses of the latest fashion, dress and undress caps, turbans.

Hardware, jewellery, plate, perfumery, fishing tackle.

Prints of droll and political characters.

IN TINS, or glass preserving jars, at a special discount for quantity:

Ox cheek and vegetables

Bristol tripe and onions

Stewed steaks

Mixed collops

Stewed kidneys

Irish stew

Haricot mutton

Oxford sausages

SOUPS: Oxtail, Julienne, hare, green pea, mulligatawny, Potted

meats, anchovy paste, Patum Peperium or Gentleman's relish.
Sardines, salmon, kippered herring, lobsters in brine, finnon
haddocks, codfish.
Green peas, mushrooms, carrots, turnips, sage & onions
Lee & Perrins Worcester Sauce
A large pack of hounds
Two very elegant suits of mourning

But imperial trade is not all carraway wafers and mixed
pickle. The supercargo knows that on board the *Blencathra*
there is also steel for bridges and railway lines, wagons for
trains, cranes, Maxim guns, rotary pumps, steam donkey-
engines, lamp-posts, nails, barbed wire, plough-shares, and
water tanks. For good or ill, India, like a hundred other places,
is being nurtured towards her future. The sailors are taming
the land as well.

thirteen

Valiant Gunner
1914

1 Main Identification Matthew Lawe in World War 1.

2 Locale Dardanelles, North Sea, Channel, Western Approaches. There are four 'set pieces' available for this section, in any or all of which Lawe may be engaged. They are:

(a) The Gallipoli landings, April 1915
(b) The Battle of Jutland, May 1916
(c) Q-ships against the U-boats, 1917
(d) The storming of Zeebrugge, April 1918

Jutland, the last great 'Grand Fleet' action which, though indecisive, kept the German fleet virtually at anchor for the rest of the war, must make its appearance, if only for the massive strength deployed by either side. The tally, never matched before or since, was:

Britain:	28	battleships		
	9	battle-cruisers		
	33	cruisers		
	79	destroyers	Total:	149

Germany:	22	battleships		
	5	battle-cruisers		
	11	cruisers		
	72	destroyers	Total:	110

But the most promising areas for a story are the last two items, (c) and (d). The 'Q-ship' was invented to counter the 'unrestricted submarine warfare' which played havoc with our supplies coming across the Atlantic or circling our coasts. It was a small armed merchantman or trawler which cruised about in any likely area looking innocent. With its gun-ports masked by false bulwarks and wooden flaps, it waited for a U-boat to surface, to look at it, to draw closer and order the crew to abandon ship.

Specially trained 'panic parties' tumbled into the boats, and the U-boat prepared to come alongside. Then the flaps crashed down, the guns rolled out, and the Q-ship's real naval crew opened fire on the U-boat. For a thoroughly hare-brained scheme comparable to the over-rated SAS forays of World War II, it was an effective counter-stroke to a desperate menace. The Germans called it 'treacherous'.

Zeebrugge (near Ostend) was a partially successful attempt to knock out the heavily armed 'mole' (breakwater) which guarded the Bruges canal, an important U-boat lair in Ger-

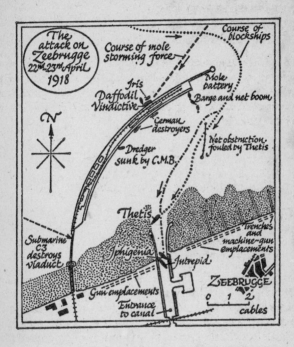

The attack on Zeebrugge 22nd 23rd April 1918

Course of blockships

Course of mole storming force

Iris
Daffodil
Vindictive

Mole battery

Barge and net boom

N

German destroyers

Net obstruction fouled by Thetis

Dredger sunk by C.M.B.

Thetis

Submarine C3 destroys viaduct

Iphigenia

Intrepid

Trenches and machine-gun emplacements

ZEEBRUGGE

0   1   2

Gun emplacements

Entrance to canal

cables

man-occupied Belgium. It involved sinking blockships (three old cruisers) at the canal entrance, but before this could be done the mole had to be stormed and its guns spiked.

The 'storm-troops' were landed from the cruiser *Vindictive* and two Mersey fery-boarts, the *Iris* and the *Daffodil*. The operation was directed by Vice-Admiral ROGER KEYES, commander of the 'Dover Patrol', and as an early 'commando' raid was valiantly fought; but the blockships were not properly

183

positioned, and Zeebrugge was a nine-day nuisance rather than a winner.

In all these actions, as in World War II, Matthew Lawe can say, with Shakespeare's citizen-soldiers, 'We are but warriors for the working day'—amateurs, volunteers, whose favourite song is—

'When this bloody war is over,
Oh, how happy we shall be! '

fourteen

Ship-Keeper to Watch-Keeper 1936-1944

1   **Main Identification** Britain's maritime decay between the two world wars, and her swift revival during the second, centred on the Battle of the Atlantic.

2   **Locale** The Clyde, English Harbour (Antigua), the Mediterranean towards Malta.

3   **Principal characters** Sailors on the scrap-heap who are needed again: an old commodore of convoys.

4   Narrative Line

The sad decay of a great maritime nation during the Thirties is reflected in Matthew Lawe's position in 1936. He is the ship-keeper in charge of a dozen old freighters, a small segment of the hundreds of rusting hulks laid up in Gare Loch on the Clyde, and in scores of other ports all round the coast of Britain. It seems that the ships are there for ever; they are not even worth towing to the breaker's yard, with the price of scrap-metal at rock bottom; and the thousands of sailors who used to man them are also permanently beached.

Sea commerce seems to have died, in the wake of the stock-market crash and the countless millions of unemployed; the world which Britain has done so much to open up, by trade, now makes its own goods, or goes without; at any rate, it no longer wants ours. Lawe has seen other lean times, other decades of neglect, but this is the worst of all.

British ships, if they sail at all, sail under sad flags of convenience: Greek, Liberian, Panamanian. A British ex-captain with an Extra Master's ticket is glad of a third mate's berth; the third mate would sign on as a bosun; and a man like Lawe spends his time listening to the suck and swallow of the tide, making tea, feeding his cats, filling in the new-fangled 'pools coupons' and shuffling to and fro across a gangplank which will never be raised again.

While he kills time—the saddest phrase in the language—Matthew day-dreams of the past. He should have seen this coming: his last voyage, in 1934, had been a sickening foretaste of this derelict world. It had taken him to the Caribbean, and to English Harbour in Antigua, in a miserable flea-bitten freighter on her last legs, forlornly island-hopping until the sea, the rats, or the owners sounded her knell.

He had never been here before, but at Burnham Thorpe Nelson had talked of it with active loathing, and as Lawe explored it, it had come alive for the second time. Or rather, it had come dead.

For English Harbour seemed to stand for England now: a ghost haunted by the past, haunted by a thousand sailors and their duels, floggings, brawls, boastings, couplings—and their deaths by wounds, disease, hanging.

Lawe's eye had traced the ruins of Nelson's Dockyard, once resounding to the clang and clamour of shipwrights, armourers, and master carpenters, men who could keep a ship seaworthy and sound, for a two- or even three-year cruise. Here was the capstan-house for heaving down (careening, they had called it in the old days), here were the sail-lockers, the mast-house, the mast-pond where they soaked the spars, the cauldrons for boiling pitch, the great catchment cisterns for saving precious rain water—all now deserted, all derelict, all given over to the ghosts.

For here, above all, Nelson, captain of the *Boreas,* still walked and wandered: fuming at the heat, the mosquitoes, the stench of a polluted harbour, and the pompous inanities of Prince William Henry. Here he had reprieved a wild-eyed deserter, a short hour before the man should have dangled from his own yard-arm ; here he had dreamed of his courtship, on an island a few score miles to the westward.

Here, after a stern chase across an entire ocean, six months before Trafalgar, he had missed Admiral Villeneuve by a few days, cursed his luck, put *Victory* onto the scent again, and brought him at last to bay.

There were sea-ghosts, and there were land-ghosts. Above the entrance to English Harbour, on Shirley Heights, these lay as thick as resurrection morning. This also was England in decline. There were old gun-emplacements, ruined cisterns, barracks and parade grounds which once had been swarming with red coats in the sunshine. There was a decaying, desolate cemetery, clinging to the steep hillside, strewn with the tumbling graves of English garrison troops, and their wives and their children, victims of colonial greatness, victims of the dreaded tropical killers— dysentry, yellow fever, sunstroke, foul distempers.

Here, on a single weathered gravestone, was summed up the end of a man's dream—a memorial to the 20-year-old wife of an artillery captain, 'Suddenly removed', with a lower, forlorn grace-note: 'Also Anne and Thomas their dear children, who departed this life in infancy, Mingle here with their kindred dust 20th October, 1851.'

Down from Shirley Heights at dusk, down once more to the harbour of 1934 and his disgraceful ship, Lawe had found the mate being carried back on board by grinning Negroes, while an old woman with an old guitar whined over and over again the catch-line of a sardonic, valedictory, malevolent calypso—

'Let him climb de riggin'
Like he climbed me.'

It was no wonder that the mate had been drunk. At first light he had shambled on deck in his under-pants, roused out the crew, and announced:

'We sail for home next week. Greenock in ballast, *if* we get there.'

'What then?'

'Then we pay off.'

They had staggered home, leaking oil and bilge-slops like a beggar with a running sore. Had that disgusting trail of flotsam marked Matthew Lawe's last voyage? The Captain, also bound for the scrap-heap, had shot himself as soon as they entered the chops of the Channel. Was this the end of *all* sailors?

It is not. Within five years of that degrading retreat, and its shoddy sequel, sailors are at a premium again. It is war! The shipwrights and the armourers are all working at full stretch; old ships are coming out of hiding, new keels are being laid as fast as the steel and the rivets can be forged. Sailors *must* be found to man them . . . It is war: Britain's huge maritime effort has to be fuelled, with blood and iron, at any cost.

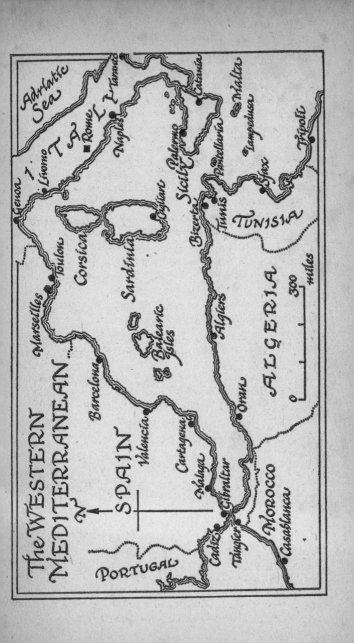

The WESTERN
MEDITERRANEAN

Adriatic
Sea

PORTUGAL

SPAIN

Genoa
Livorno
ITALY
Rome
Naples

Marseilles
Toulon

Corsica

Sardinia

Cagliari

Balearic
Isles

Barcelona

Valencia

Cartagena

Malaga
Gibraltar

Cadiz
Tangier

MOROCCO

Casablanca

Oran

ALGERIA

Algiers

Bizerta

Tunis

TUNISIA

Sfax

Tripoli

Lampedusa

Pantellaria

Palermo

Sicily

Catania

Malta

300 miles
0

Lawe, with only one bitter glance backwards, the last of so many—why does it always take a war to give us back our manhood?—crosses the rickety gangplank and goes off to fight.

Two episodes are enough to cover World War II: (a) a Malta convoy in 1942, and (b) the Normandy landings in 1944. As recent history, they need only be presented in snapshots.

(a) The commodore's ship of a small Malta convoy, heavily bombed, out of action, with most of her crew killed, drifts miles off her course during a black night, and into a minefield off the Tunisian coast. She cannot be reached by any rescue vessels, and she is bound to sink, anyway. At daylight, a cruiser on the far horizon closes to investigate, and might be in danger herself. The commodore, a dug-out admiral of seventy, signals: 'We are not worth a cruiser. Suggest goodbye.' The reply is: 'Concur. God bless you.' After that they are alone.

After days of drifting among the shoals south of Cape Bon, they are still afloat. No one bothers them: the main battle is far away. Then a westerly breeze gets up, and increases. They have no engines, but they might be able to steer by tiller-ropes from aft. Matthew Lawe suggests to the commodore: 'Sir, how about "Down funnel, up sail"?'

On the mast and the derricks they rig an absurd collection of sheets, blankets, carpets, old clothes, and some black-out curtains destined for the Flag Officer, Malta. But it moves the ship! Square-rigged under all plain sail, they wallow down wind towards Malta, and safety.

(b) On the first night of Operation Overlord, the Allied landing in Normandy, the might of the British-American fleets is launched against the coast of Fortress Europe. There are 5000 ships, the greatest armada ever to sail. Matthew Lawe's share in it comes some hours later, and is a humble one: he forms part of the 'scuttle party' on board one of the rotten old ships which he has been 'keeping' up at Gare Loch.

She is the same kind of 'expendable' as the fire-ships of Francis Drake against the Armada: destined to be sunk, to form one of the pivots of Mulberry Harbour, the floating gang-way from the sea to the shore.

It is skilfully, exactly, and properly done. The old ship, nudged into position by RNVR motor-launches, settles and sinks. Water enters every part of her, throttling the passage-ways. Finally choked by the sea, she becomes an instant link in the Mulberry causeway. So she is some good after all.

Mulberry, and the other harbours, put a million men ashore on the enemy coast in twenty-eight days: doing what the Spanish Armada failed to do, and Napoleon never dared, and Hitler cancelled for prudent reasons.

Back in harbour on the blessed coast of Sussex, some of Lawe's shipmates are inclined to sneer at, or be jealous of, the vast American fleet.

'It's their turn now,' Lawe says.

'How do you mean?'

'Just you wait and see.'

Peerless sea-power has passed to other, willing hands.

fifteen

The Good Acquittal 1978

1 **Main Identification** The end of the story, in the poetry of death.

2 **Locale** Southampton: the St Lawrence Seaway from Quebec to the locks of Sault Ste Marie on Lake Superior.

3 **Principal characters** Striking British dock workers, young Canadian sailors, and a girl with a guitar.

4 **Narrative Line**

The day before the liner *Queen Elizabeth 2*, already delayed by strikes, is due to sail on a world cruise of great financial importance, there is a 'demarcation dispute' which has a farcical beginning and a catastrophic end. Matthew Lawe, a steward in the tourist section, is the continuing culprit in the final setback, which involves the installation of a new rubber pad to cushion a raised lavatory seat.

This work of art requires the representatives of seven unions to bring it to full flower: a painter, a polisher, a plumber, a carpenter (to drive in the screws, and insert a wooden backing for the pad), a shipwright (for drilling a hole in 'the main structure'), a sweeper to keep on tidying up, and an electrician to replace a light bulb, not quite strong enough to work by.

Lawe, who wants to get on with his work, spread among five other cabins, keeps trying to help, and excites horror and then resentment among the union men. First he brushes up some of the preliminary mess from his spotless carpets, and is told to stop it: then he makes as if to change the light-bulb, inducing agony among everyone present, for this is the work of the strongest union of all.

'Sabotage, that's what it is!' the ETU man tells him. 'You're not assimilated, brother, that's your trouble.'

'Come off it, for Christ's sake!' Lawe says, ashamed and embarrassed. 'I'm used to turning my hand to anything.'

'Not while *I'm* around.' A baleful stare enforces the words. 'Whose side are you on, anyway? Just you watch it—because I'm watching *you*! '

Lawe's final gaffe is to raise the lavatory seat, to test the new installation. 'But you've put it on upside down! Look.' He shows them how it is no good, causing first a menacing silence and then uproar. Whoever it is who should point out such faults (if they *are* faults), it is *not* a sailor. Moving or adjusting ship's furniture is a Transport & General Workers' Union job, anyway.

It is yet another slap in the face for the working class. 'One out, all out!' is the cry, and they all troop off the ship for the last time, leaving her idle and useless. The cruise has to be cancelled, at an enormous loss of money and prestige (hailed as 'a brilliant victory' by a TUC boss next day), and Lawe is out of work, and a marked man with the Seamen's Union as well. Weary and dispirited (can it be that the British are now trying to kill off their own ships?), he wanders for a while. Then he finds himself in Canada, with a steward's job again—but a steward on a bulk oil-carrier traversing the length of the St Lawrence Seaway, from Quebec to the lake-head at Fort William, on Lake Superior.

Fort William lies in the very heart of the North American continent: 2,250 miles from the sea, the furthest penetration of the land ever made by sailors. To build this big-ship fairway and extend this frontier, islands have been sliced in half or removed altogether, rivers dammed or diverted, whole villages set aside, channels blasted out of the living rock.

It is a marvellous journey, through a noble landscape: from Quebec with all its old memories, past little landing places and big industrial towns; through forests of trees and forests of cranes, across huge lakes whose furthest margins cannot be seen. Sometimes they are in American waters, sometimes in Canadian; but all the time the ship, and Matthew Lawe, are voyaging deeper towards peace, towards refuge, towards some secret haven which exists on neither map nor chart until it is reached by the fortunate traveller.

Lawe, an industrious historian, logs the ships of twenty-nine nations from Liberia to China, Iceland to Israel, South Africa to the strangest lair of all, landlocked Switzerland. He passes one which gives him an aching flavour of the past: a small salty ship whose port of registration reads 'Bideford,

England'. *Men of Bideford in Devon* . . . It is the song of the little *Revenge*, the fight of the one against the fifty-three . . . 'At Flores in the Azores, Sir Richard Grenville lay' . . . And died, in 1591, before some sailors were even born.

Matthew Lawe gazes longingly as the little ship of Bideford, with her proud label, disappears down channel. Perhaps the blood of Devon is still there, after all.

On board his own ship, things are less happy than this dream and less handsome than the land. Lawe is a figure of fun to the young crew: an old sailor past his best, good for nothing but passing plates, washing up, sweeping up, and ditching the galley trash at the day's end—the lads know all this, and are merciless in their taunting.

'You're past it, daddy-oh!' they keep on saying, as if Matthew were in truth the despised father figure, whom no younger generation can either honour, or tolerate, or leave in peace.

At the locks of Sault Ste Marie, 250 miles from the lakehead and their journey's end, they have an idle day as the ship discharges some of her oil. The crew's baiting reaches a cruel level. Now it extends not only to old Matthew Lawe, but to Britain as well—the proud imperial totem toppled to the ground. All that Matthew can answer is: 'Maybe we *are* past it, now. But we had a good run.'

' "Run" is right!' his chief tormenter says spitefully. 'Remember Dunkirk?'

He would like to counter with 'Remember Trafalgar? Remember Drake at Ushant?' But what's the point? . . . Standing on the upper deck, Lawe looks round him, past the mocking faces to some of the dividends of a sea-borne Admiralty. A convoy of ships in the heart of a continent. A thriving city which was once an Indian canoe-portage. Aircraft crossing a continent: helicopters air-lifting whole prefabricated towns to be set down in the Canadian northland, far beyond the snowline, where only the barren ice used to reign. And this is only one country! Look round the circle of the globe. It all began with the sea, with sailors; and England was the cradle, the layer of the keels.

'We had a good run,' he says again. 'Maybe I *am* past it, like

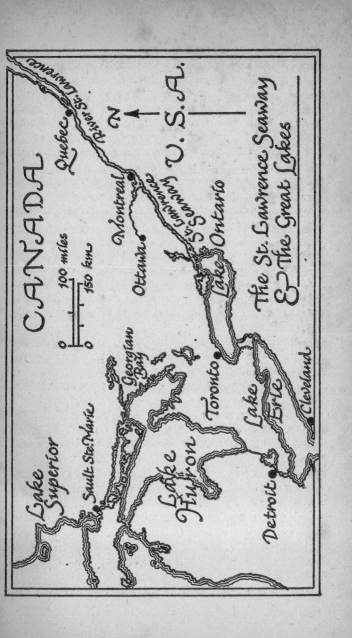

CANADA

100 miles

150 km

Lake Superior

Sault Ste. Marie

Lake Huron

Georgian Bay

Detroit

Lake Erie

Cleveland

Toronto

Lake Ontario

Ottawa

Montreal

Quebec

River St. Lawrence

St. Lawrence Seaway

N

U.S.A.

The St. Lawrence Seaway & The Great Lakes

England. But by God, by God——'

He does not know how to finish. Followed by jeering laughter he makes his escape, and goes ashore.

He escapes to pastoral, idyllic peace, and a languorous dream of contentment summed up in one comely girl. Sitting by the roadside, backed by a grove of pine trees, within sight of his ship in the harbour below, Lawe falls asleep, and is awakened by the notes of a guitar. There is a girl perched on the bank by his side: a gravely pretty girl, with a rucksack and a guitar, and the washed-out blue jeans and halter top which customarily go with this outfit.

She is smiling, as she repeats on the guitar the grace-notes which had woken him. 'I wanted someone to talk to,' she says. 'Who are you and why are you asleep? Are you a wanderer?'

With her ravishing smile, long coltish body, and open innocence, she is the very wine of youth. There is a swift sensual message, but it cannot be strong: this guileless child is not the meat of conquest. When, still collecting his wits, he asks her to sing, she answers, 'Some of my songs are sleep-songs. You want a wake-up song?'

She sings him a wake-up song, and then he answers her first question. Yes, he is a sort of wanderer. He is a sailor. That's his ship down there.

'Is it hard? Is there a lot to learn?'

He quotes:

'Learn of the little nautilus to sail,
Spread the thin oar, and catch the driving gale.'

'What's a little nautilus?'

'One of our ancestors.' He elaborates. 'It's also called an argonaut. It's like a clever oyster. It has a thin shell, like a boat, but thin like paper. It can lie still, or it can paddle along, or it can come to the surface and raise its little arms and sail with the wind.'

'Why did you say, one of our ancestors?'

'They're all our ancestors.'

She is avid to learn, whether of life or its ancient mechanics. 'Tell me some more ancestors. Teach me something for today.'

He tell her something of the irresistible continuity of life. Even the little nautilus had its own ancestors; indeed, it was deeply sophisticated, compared with the dark-fronded seaweed from which it came.

From there grew the great amorphous molluscs and the armoured sea-saurians, the exploring crab-like creatures at the margin of the tide: the ravaging land-animals, the apes and then the people who took possession of the earth, the very wood for their wandering ships: the furnace which overwhelmed Pepys's London, the salt-horse for Nelson's sailors, the mouldy caribou-skins which kept Franklin's men alive in the howling wilderness of the northland: the mineral wealth, the oil and coal, the wish and the means to destroy by explosive violence, the whole towering edifice of our civilization, even the St Lawrence Seaway lying below them now.

All these have sprung from some blind, minute, determined embryo which, moving sluggishly with the wash of the tides, ten million years before old Drake first scorched the Spanish Main, set our pattern and formed our fearful lives.

The girl listens attentively, though perhaps the lesson has been rather too tough for a lazy summer afternoon. At the end she sighs.

'So everything is really very old.'

'Very old.'

'Does that mean that we can't win?'

'It means that we can.'

She smiles brilliantly. 'Then I'll sing you an old song! But with *lots* of hope!'

Her song is borrowed from Robert Burns ('But I can't do the funny accent'), and is the sad and yearning *A Red Red Rose*, fitted to her music like one loving hand to another:

'My love is like a red red rose
That's newly sprung in June:
My love is like the melody
That's sweetly played in tune.

As fair are you, my bonnie lass,
So deep in love am I:

And I will love you still, my dear,
Till all the seas run dry.

Till all the seas run dry, my dear,
And the rocks melt with the sun,
I will love you still, my dear,
While the sands of life shall run.

So fare you well, my only love,
And fare you well awhile.
And I will come again, my love
Though it were ten thousand mile!'

Matthew joins softly in the last words. Then he repeats: 'Till all the seas run dry.' They look into each other's eyes, and she says: 'Matt! You're nearly crying!'

'It's so sad.'

'It's triumphant!'

The moment is shattered by hideous alarm and uproar from the ship below them, Matthew's ship. A vile-sounding hooter signals danger, and keeps on doing so, in endless barks of fear. A great gulp of black smoke erupts, and then drifts up the hillside. The oil carrier is on fire.

With a brief word of farewell to a loving friend, Matthew races downhill. When he arrives, panting, it is obvious what has happened. During de-oiling, a hose on board has leaked, and then caught fire. Half the upper deck is already engulfed in flame.

There is by now a wall of fire between the main blaze and a deck-barrier leading to an open hatch. The door in it *must* be closed: otherwise the ship will go up like a torch, and take half Sault Ste Marie with it.

This time, Matthew Lawe does not hesitate. All he can remember is that first fatal act of cowardice on Drake's fire-ship, and the voice of a man, horribly burned, whispering and snarling down the centuries: 'Matthew! You run away! Curse you for a coward! *Do you want to live forever?*'

It is enough. Charging through a fiery hell, he darts forward and slams shut the door leading to the hatchway. Then he drops senseless, his body roasted, his clothes in smouldering rags. When the fire subsides, he is dragged out, and laid gently on the dockside.

Under the evening sky, waiting for an ambulance which can only arrive too late, his chief tormentor apologizes. 'We didn't mean it, Matt. About you being too old. Christ, you've proved it now!'

'It doesn't matter, lad.' Dying, the West Country speech is an exhausted growl. 'Maybe you were right.' He looks up at the darkening sky, yet what he sees is the great line of ships marching through the locks, the giant seaway which suports the probing aircraft, all the evidence of sailors still voyaging and discovering.

Now at last he knows how to finish. 'Maybe I *am* past it. But by God I have sons!'

A *possible* last few paragraphs: From the hospital, a doctor is telephoning the police with his report on the death. It is a run-of-the-mill accident, though not less agonizing on that account, multiple burns and shock. There was never any chance of saving Matthew Lawe.

But: 'Is this the only casualty? I mean, have I got the right man? The lads that brought him in said something about an old steward.'

'How old is he, then?'

'Twenty-two or three. And strong as an ox.' But the doctor has plenty of other things to do. 'Mind you, he was under a sheet already . . . I guess someone will sort it out.'

Nicholas Monsarrat
The White Rajah £1.75

Fierce battles, voluptuous dances and barbaric tortures breathtakingly mingle in this tremendous historical novel. Makassang – an island in the tropic Java Sea, at once savage and splendid, red with the stain of blood and the glow of rubies . . .

'A fine swashbuckler by an accomplished storyteller'
NEW YORK POST

The Ship That Died of Shame, and Other Stories 90p

The story of a gunboat, dishonoured by smugglers, which took its revenge in a raging sea . . . the tale of a honeymoon-turned-manhunt when an ex-commando employs the tricks of his trade, in 'Licensed to Kill' . . . and in 'The Thousand Islands Snatch' we read of days of terror for a beautiful girl – hostage for a fortune . . .

The Kappillan of Malta £2.25

In the figure of Father Salvatore, who staunches the blood of the wounded and runs his own church in the ancient catacombs, Monsarrat has created one of the most memorable characters of postwar fiction' DAILY EXPRESS

Patrick Raymond
The White War £1.25

July 1944 – the end of the war in sight – Allied troops prepare to launch the final crushing attack on Nazi Germany. When a Polish squadron arrives to share the disciplined quarters of a Yorkshire bomber station, a strange time begins – a time of duels, disappearances, a maze of riddles for Station Commander David Armstrong, and at the root of it all the enigmatic Polish pilot Prince Karol, a man torn between vengeance and duty, driven by the memory of atrocity . . .

Wilbur Smith
A Sparrow Falls £1.75

From the trenches of France in the First World War, through the violence of the Johannesburg strikes of the 1920s, the last chapter of Sean Courtney's epic life takes him home to the serene splendours of the African wilderness.

'A man's world, where hate can swell like biceps and frontiers beckon as seductively as a woman ... Wilbur Smith writes as forcefully as his tough characters act'

Cry Wolf £1.75

Two men, one girl and a batch of decrepit armoured cars ... running the gauntlet of an Ethiopia in the grip of the Wolf of Rome. 'Mussolini has all the guns, aircraft and armour he needs. The jolly old Ethiop has a few ancient rifles and a lot of two-handed swords ... It should be a close match!'

'Another cracker ... Africa, arms dealing, armoured cars, strong men with stronger women all combine beautifully for real entertainment' DAILY MIRROR

Eagle in the Sky £1.50

In Israel's nerve-stretching struggle for survival, David Morgan's brilliance as a Mirage pilot is his passport to Debra's love. But terrorism and tragedy spawned by the violence that drew them together threaten to tear them apart. From savage air-fights over the desert to hand-to-hand conflicts on a South African game reserve, this unforgettable story blends intense excitement with a tender, sensual love ...

The Eye of the Tiger £1.50

'A blood-and-cyclone story ... action follows action, menace mingles with violence and horror and Harry Fletcher with blonde and brunette ... Mystery is piled on mystery ... death on brutal death as Fletcher hunts the Mozambique Channel for the ocean's unknown treasure. A tale to delight the millions of addicts of the gutsy adventure story' SUNDAY EXPRESS

Colin Forbes
The Palermo Ambush £1.50

'As inventive as Alistair Maclean . . . this desperate adventure
involving the Mafia and the Nazis is as ingenious and
blood-bolstered as they come' SUNDAY TIMES

'A projectile of a thriller that whooshes through twenty-four
hours of maximum tension . . . with a thunderous finale that outdoes
even the chase that precedes it' NEW YORK TIMES

The Stone Leopard £1.50

'A real cracker ripping through Europe. The heart of the story is a plot
which could destroy the Western world. The action is based on
the threat to a President of France from a mysterious and vicious
resistance leader from the past. Hard, bitter and compulsive;
and the ending . . . Wow I' DAILY MIRROR

Year of the Golden Ape £1.50

Racing at Concorde speed from the Middle East through Europe to
California, a ruthless battle of wits reaches unparalleled tension as
mercenaries hijack a giant British tanker, arm her with a
plutonium bomb and sail into San Francisco harbour.

Target Five £1.50

No quarter is asked or given when a top Russian oceanographer
defects across the Arctic icefields with plans of their submarine
network. The Americans send in dog teams and an unconventional
trio under Anglo-Canadian agent Keith Beaumont. The Russians
use everything they have in an increasingly bloody life-or-death
struggle to win him back . . .

Dick Francis

'If he ever opens a detective agency or joins the British intelligence service, some of the supersleuths of literature will have to tend to their laurels' TIME MAGAZINE

Enquiry £1.25

'All the elements that Dick Francis handles so superbly! Horse-racing so vibrantly portrayed that it fairly gallops from the pages ... filled with suspense, high drama and the bristling hatred of revenge' NEW YORK TIMES

Rat Race £1.25

'Top-class novel about a taxi-plane pilot involved in a fraud scheme ... exciting (there is an in-the-air rescue sequence that would make you really and truly angry if you had to put it down) and a splendid read' THE TIMES

Smokescreen £1.25

'Certainly his best thriller ... sets the scene of devilry in South Africa. Truly exciting' EVENING STANDARD

For Kicks £1.25

'An absolute beauty ... the detection is ingenious and detailed ; the gimmick is a fine one ; and the background of life among horses and trainers and stable lads (and criminals) is so real you can smell and taste it. As a puzzle, as a thriller, it's a winner' NEW YORK TIMES

Bonecrack £1.25

'This time it's a rich monomaniac who will stop at nothing, not blackmail, not murder, to ensure his son will be a crack Derby jockey. A classic entry with a fine turn of speed' EVENING STANDARD

Selected Bestsellers

☐	**Gone with the Wind**	Margaret Mitchell	£2.95p
☐	**Robert Morley's Book of Worries**	Robert Morley	£1.50p
☐	**The Totem**	David Morrell	£1.25p
☐	**The Alternative Holiday Catalogue**	edited by Harriet Peacock	£1.95p
☐	**The Pan Book of Card Games**	Hubert Phillips	£1.50p
☐	**The New Small Garden**	C. E. Lucas Phillips	£2.50p
☐	**Food for All the Family**	Magnus Pyke	£1.50p
☐	**Everything Your Doctor Would Tell You If He Had the Time**	Claire Rayner	£4.95p
☐	**Rage of Angels**	Sidney Sheldon	£1.75p
☐	**A Town Like Alice**	Nevil Shute	£1.50p
☐	**Just Off for the Weekend**	John Slater	£2.50p
☐	**A Falcon Flies**	Wilbur Smith	£1.95p
☐	**The Deep Well at Noon**	Jessica Stirling	£1.75p
☐	**The Eighth Dwarf**	Ross Thomas	£1.25p
☐	**The Music Makers**	E. V. Thompson	£1.50p
☐	**The Third Wave**	Alvin Toffler	£1.95p
☐	**Auberon Waugh's Yearbook**	Auberon Waugh	£1.95p
☐	**The Flier's Handbook**		£4.95p

All these books are available at your local bookshop or newsagent, or can be ordered direct from the publisher. Indicate the number of copies required and fill in the form below

Name_____
(block letters please)

Address_____

_____

Send to Pan Books (CS Department), Cavaye Place, London SW10 9PG
Please enclose remittance to the value of the cover price plus:
25p for the first book plus 10p per copy for each additional book ordered
to a maximum charge of £1.05 to cover postage and packing
Applicable only in the UK
While every effort is made to keep prices low, it is sometimes
necessary to increase prices at short notice. Pan Books reserve
the right to show on covers and charge new retail prices which
may differ from those advertised in the text or elsewhere